Real-time Computer Control

Prentice Hall International Series
in Systems and Control Engineering

M.J. Grimble, Series Editor

BANKS, S.P., *Control Systems Engineering: modelling and simulation, control theory and microprocessor implementation*
BANKS, S.P., *Mathematical Theories of Nonlinear Systems*
BENNETT, S., *Real-time Computer Control: an introduction*
CEGRELL, T., *Power Systems Control*
COOK, P.A., *Nonlinear Dynamical Systems*
PATTON, R., CLARK, R.N., FRANK, P.M. (editors), *Fault Diagnosis in Dynamic Systems*
SÖDERSTRÖM, T., STOICA, P., *System Identification*

Real-time Computer Control

STUART BENNETT

Department of Control Engineering,
University of Sheffield, UK

PRENTICE HALL

NEW YORK • LONDON • TORONTO • SYDNEY • TOKYO

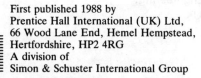

First published 1988 by
Prentice Hall International (UK) Ltd,
66 Wood Lane End, Hemel Hempstead,
Hertfordshire, HP2 4RG
A division of
Simon & Schuster International Group

© 1988 Prentice Hall International (UK) Ltd

All rights reserved. No part of this publication may be
reproduced, stored in a retrieval system, or transmitted,
in any form or by any means, electronic, mechanical,
photocopying, recording or otherwise, without the
prior permission, in writing, from the publisher.
For permission within the United States of America
contact Prentice Hall Inc., Englewood Cliffs, NJ 07632

Printed and bound in Great Britain at
The University Press, Cambridge

Library of Congress Cataloging-in-Publication Data

Bennett, S. (Stuart)
  Real-time computer control.

  (Prentice-Hall International series in systems and
control engineering)
  Bibliography: p.
  Includes index.
  1. Digital control systems. 2. Real-time data
processsing. I. Title. II. Series.
TJ223.M53B46 1988   629.8′95   87–29274
ISBN 0–13–762485–9

---

1 2 3 4 5   92 91 90 89 88

ISBN 0-13-762485-9
ISBN 0-13-762501-4 PBK

**In memory of the people of Sheffield
who by public subscription founded a University**

Contents

Preface	xiii
Acknowledgements	xv

1 Introduction to Real-time Systems — 1

1.1	Historical Background	1
1.2	Elements of a Computer Control System	3
1.3	Classification of Real-time Systems	8
	1.3.1 Clock-based systems	10
	1.3.2 Sensor-based systems	10
	1.3.3 Interactive systems	11
1.4	Real-time Systems – a Definition	11
1.5	Classification of Programs	14
1.6	Summary	16
	Exercises	16
	References and Bibliography	17

2 Concepts of Computer Control — 19

2.1	Introduction	19
2.2	Sequence Control	21
2.3	Loop Control (Direct Digital Control)	26
2.4	Supervisory Control	33
2.5	Human or Man–Machine Interface (MMI)	36
2.6	The Control Engineer	37
2.7	Centralized Computer Control	38
2.8	Hierarchical Systems	39
2.9	Distributed Systems	43
2.10	Economics of Computer Control Systems	45
	Exercises	46
	References and Bibliography	46

3 Computer Hardware Requirements for Real-time Applications — 48

3.1	Introduction	48
3.2	General Purpose Computer	48
	3.2.1 Central processing unit	48
	3.2.2 Storage	52
	3.2.3 Input and output	53
	3.2.4 Bus structure	53

	3.3	Process-Related Interfaces	54
		3.3.1 Digital signal interfaces	55
		3.3.2 Pulse interfaces	59
		3.3.3 Analog interfaces	61
		3.3.4 Real-time clock	63
	3.4	Data Transfer Techniques: Polling	64
	3.5	Data Transfer Techniques: Interrupts	67
		3.5.1 Saving and restoring registers	68
		3.5.2 Interrupt input mechanisms	69
		3.5.3 Interrupt response mechanisms	71
		3.5.4 Hardware vectored interrupts	74
		3.5.5 Interrupt response vector	80
		3.5.6 Multilevel interrupts	84
	3.6	Comparison of Data Transfer Techniques	85
		3.6.1 Direct memory access	87
	3.7	Communications	87
		3.7.1 Asynchronous and synchronous transmission techniques	88
		3.7.2 Local and wide area networks	92
	3.8	Standard Interfaces	94
		Exercises	95
		References and Bibliography	97

4 DDC Control Algorithms and their Implementation — 99

4.1	Introduction	99
4.2	The PID Control Algorithm: the Basic Algorithm	100
4.3	Implementing the Ideal PID Controller	101
4.4	Timing	103
4.5	Alternative Forms of the PID Algorithm	106
	4.5.1 Bumpless transfer	106
	4.5.2 Saturation	109
	4.5.3 Noise	115
	4.5.4 Improved forms of algorithm for integral and derivative calculation	117
4.6	Tuning and Choice of Sample Interval	118
4.7	Implementation of Controller Designs Based on Plant Models	120
	4.7.1 Direct methods	120
4.8	The PID Controller in z-transform Form	123
4.9	Summary	124
	Exercises	125
	References and Bibliography	127

5 Design of Real-time Systems — 129

5.1	General Approach	129
5.2	Specification Document	133
5.3	Preliminary Design	135
	5.3.1 Hardware design	135

	5.3.2	Software design	137
5.4		Single Program Approach	138
5.5		Foreground/Background System	140
5.6		Multi-tasking Approach	144
5.7		General Approach to Real-time Software Design	145
5.8		MASCOT	150
	5.8.1	Activity	150
	5.8.2	Communication	150
	5.8.3	Channels	151
	5.8.4	Pools	151
	5.8.5	Synchronization	152
5.9		Example of Preliminary Design	154
5.10		Detailed Design: Module Subdivision	157
5.11		Design Review	160
5.12		The MASCOT System	162
5.13		Structured Development for Real-time Systems	162
	5.13.1	Data transformation	163
	5.13.2	Control transformations	165
	5.13.3	Prompts	165
	5.13.4	Summary of the method	167
	5.13.5	Building the model	167
5.14		Summary	172
		Exercises	174
		References and Bibliography	174

6 Operating Systems   176

6.1		Introduction	176
6.2		Single-task or Single-job Operating System	179
	6.2.1	CCP direct commands	180
	6.2.2	Basic disk operating system	182
6.3		Simple Foreground/Background Operating System	189
	6.3.1	General foreground/background monitors	191
6.4		Real-time Multi-tasking Operating Systems	192
6.5		Task Management	195
	6.5.1	Task states	195
	6.5.2	Task descriptor	197
6.6		Task Dispatch and Scheduling	201
	6.6.1	Priority levels	201
	6.6.2	Interrupt level	203
	6.6.3	Clock level	204
	6.6.4	Cyclic tasks	204
	6.6.5	Delay tasks	208
	6.6.6	Base level	208
	6.6.7	System commands which change task status	209
	6.6.8	Dispatcher: search for work	211
	6.6.9	Deadlock	214
6.7		Memory Management	215

6.8	Code Sharing	219
	6.8.1 Serially reusable code	220
	6.8.2 Reentrant code	220
6.9	Input/Output Sub-system (IOSS)	222
	6.9.1 Example of an IOSS	226
	6.9.2 Output to printing devices	227
	6.9.3 Example of input from keyboard	228
	6.9.4 Device queues and priorities	230
6.10	Task Cooperation and Communication	230
6.11	Summary	232
	Exercises	233
	References and Bibliography	233

7 Concurrent Programming 234

7.1	Introduction	234
7.2	Concurrent Programming	235
7.3	Mutual Exclusion	237
	7.3.1 Primitives	239
	7.3.2 Condition flags	240
	7.3.3 Semaphores	243
7.4	Producer-consumer Problem	249
7.5	Monitors	258
7.6	Rendezvous	262
7.7	Summary	268
	Exercises	268
	References and Bibliography	269

8 Real-time Languages 270

8.1	Introduction	270
8.2	User Requirements	270
	8.2.1 Security	271
	8.2.2 Readability	272
	8.2.3 Flexibility	274
	8.2.4 Simplicity	280
	8.2.5 Portability	280
	8.2.6 Efficiency	280
8.3	Language Requirements and Features	281
8.4	Declarations	283
8.5	Types	284
	8.5.1 Sub-range types	285
	8.5.2 Derived types	286
	8.5.3 Structured types	287
	8.5.4 Pointers	287
8.6	Initialization	288
8.7	Constants	288

	8.8	Control Structures	289
	8.9	Scope and Visibility	291
	8.10	Modularity	294
	8.11	Independent and Separate Compilation	296
	8.12	Exception Handling	298
	8.13	Low-level and Multi-tasking Facilities	301
		Exercises	303
		References and Bibliography	303

9 Programming Languages 305

	9.1	Assembly Languages	305
	9.2	Evolution of High-level Languages	307
	9.3	BASIC	311
	9.4	FORTRAN and Pascal	312
	9.5	CORAL 66	315
	9.6	RTL/2	316
	9.7	Modula-2	318
		9.7.1 Modules	320
		9.7.2 Low-level facilities	323
		9.7.3 Concurrent programming: co-routines	324
		9.7.4 Concurrent programming: processes	326
		9.7.5 Interrupts and device-handling	328
		9.7.6 High-level multi-tasking modules	330
	9.8	Ada	337
	9.9	Application-oriented Software	338
		9.9.1 Table-driven	338
		9.9.2 Block-structured software	341
		9.9.3 Application languages	342
	9.10	CUTLASS	342
		9.10.1 General features of CUTLASS	345
		9.10.2 Data typing and bad data	347
		9.10.3 Language sub-sets	348
		9.10.4 Scope and visibility	348
		9.10.5 Summary	350
	9.11	Choice of Programming Language	350
		References and Bibliography	355

Index 359

Preface

Over the past 25 years the application of digital control to industrial processes has changed from being the exception to the commonplace. The growth in the applications of computer control has been brought about largely by the rapid advances in hardware design and the reduction in costs: this is most clearly demonstrated by the extent to which the microprocessor has become a normal component of a wide range of electronic systems.

The development of design and production techniques for the software necessary to operate computer systems has not kept pace with the advances in hardware. The problems of software design and production are well-known and there are major research and development programs aimed at remedying the position. These programs are concerned both with developing techniques for design and with the provision of software tools to support both design and production. It is now clearly understood that the creation of software for real-time systems, i.e. systems which have to respond in real-time to events in the outside world, is one of the most difficult areas of software design and production.

The difficulty faced by both engineering students and by experienced engineers is that traditional computing courses for engineers have emphasized the hardware and programing language aspects of computer systems. The languages usually taught – FORTRAN and more recently Pascal – do not have any support for real-time concurrent processing or for direct manipulation of the computer hardware. Applications used to illustrate the teaching have usually been restricted to stand-alone scientific programs. Attempts to go beyond this have required instruction in the use of assembly languages. The majority of books which deal with real-time applications and with digital control assume that the software will be written in an assembly language. They concentrate almost entirely on the techniques for coding algorithms and are therefore concerned with speed of execution, memory usage, and the effects of word length on accuracy. These are, of course, important problems in many applications; however, there are applications for which a different approach, based on the use of high-level languages is appropriate and this book attempts to address this area.

The book is intended for final year undergraduate students and practising engineers. It assumes that the reader will have some familiarity with at least one high-level language and possibly with an assembly language as well. The first part of the book (Chapters 1, 2 and 3) provides a general overview of the subject including definitions and classifications of real-time systems, computer control configurations and hardware requirements.

Chapter 4 deals with methods of implementing a controller on a digital computer. The problems are illustrated by considering the implementation of the traditional three-term (PID) controller. A brief coverage of the implementation of a controller given in z-transform notation, obtained either by the use of discrete design methods or through discretization of a continuous controller design, is given. Knowledge of z-transforms is not required in order to understand the section.

Software design techniques for real-time systems are introduced in Chapter 5. The principles underlying two methods, MASCOT and real-time structured design, are covered. Emphasis is given to dividing the software into modules and to the use of multi-tasking. The traditional approach to implementing multi-tasking – the use of a real-time operating system – is covered in Chapter 6. The general features of operating systems are introduced by describing a simple single-user, single-task, operating system (CP/M 80). The various additional requirements for real-time multi-tasking are then introduced.

In Chapter 7 the approach to supporting multi-tasking based on the use of a real-time language with minimum operating system support are considered. The basic ideas of concurrent programming, use of semaphore, signals and monitors are described. The general language requirements for real-time programming are covered in Chapter 8 and a brief comparison of a number of languages is given in Chapter 9. The examples in the book are given in several languages, but predominantly in Modula-2.

Many people have assisted in producing this book and I am grateful to all of them. Particular thanks are due to Steve White, a former colleague; to the many students who have assisted in developing my understanding both through class discussion and through the project work which they have carried out, and to the technical staff of the Department of Control Engineering, University of Sheffield. I wish also to thank the staff of Prentice Hall, in particular Glen Murray and Andrew Binnie, for their advice and support.

S.B.

Acknowledgements

The Publishers wish to thank the following for permission to reproduce extracts from published material:

Peter Peregrinus Ltd, for Figures 2.1, 2.5, 2.14, 9.29 reproduced from S. Bennett and D.A. Linkens (1984), *Real-time Computer Control* and for Figure 2.8, reproduced from S. Bennett and D.A. Linkens (1982), *Computer Control of Industrial Processes*.

The Institution of Electrical Engineers for Figures 8.4, 8.5, 8.6, 8.7 reproduced from B.S. Hoyle (1984) 'Engineering microprocessor software', *Electronics and Power*, **30**.

The American Society of Mechanical Engineers for Figure 1.1 redrawn from G.S. Brown and D.P. Campbell (1950) 'Instrument engineering: its growth and promise in process control problems', *Mechanical Engineering*, **72**, p. 124.

Ellis Horwood for Figure 7.7 reproduced from S.J. Young (1982), *Real-time Languages*.

Van Nostrand Reinhold for Figure 9.24 reproduced from D.A. Mellichamp (ed.) (1983), *Real-time Computing*.

1

Introduction to Real-time Systems

1.1 HISTORICAL BACKGROUND

The earliest proposal to use a computer operating in 'real time' as part of a control system was made in a paper by Brown and Campbell [1950]. The paper contains a diagram (see Figure 1.1) which shows a computer in both the feedback and feedforward loops. Brown and Campbell assumed that analog computing elements were the most likely to be used but they did not rule out the use of digital computing elements. The first digital computers developed specifically for real-time control were for airborne operation, and in 1954 a Digitrac digital computer was successfully used to provide an automatic flight and weapons control system.

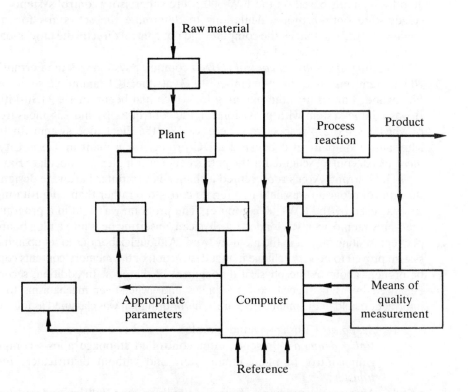

Fig. 1.1 Computer used in control of plant (redrawn from Brown and Campbell *Mechanical Engineering*, **72**, 1950).

The application of digital computers to industrial control began in the late 1950s. The initiative came, not from the process and manufacturing industries, but from the computer and electronic systems manufacturers who were looking to extend their markets and to find outlets for equipment which had failed to be adopted by the military [Williams 1983]. The first industrial installation of a computer system occurred in September 1958 when the Louisiana Power and Light Company installed a Daystrom computer system for plant monitoring at their power station in Sterling, Louisiana. This was not a control system: the honor of the first industrial computer control installation went to the Texaco Company who installed an RW-300 (Ramo-Wooldridge Company) system at their Port Arthur refinery in Texas, which achieved closed-loop control on March 15, 1959 [Anon 1959].

During 1957–8 the Monsanto Chemical Company in cooperation with the Ramo-Wooldridge Company, studied the possibility of using computer control and in October 1958 decided to implement a scheme on the ammonia plant at Luling, Louisiana. Commissioning of this plant began on January 20, 1960 and closed-loop control was achieved on April 4, 1960 after an almost complete rewrite of the control algorithm part of the program and considerable problems with noise on the measurement signals. This scheme, like the system installed by the B.F. Goodrich Company on their acrylanite plant at Calvert City, Kentucky in 1959–60, and some 40 other systems based on the RW-300, were supervisory control systems used for steady-state optimization calculations to determine the set-points for standard analog controllers; that is, the computer did not control directly the movement of the valves or other plant actuators.

The first *direct digital control (DDC)* computer system was the Ferranti Argus 200 system installed in November 1962 at the ICI ammonia-soda plant at Fleetwood, Lancashire, the planning for which had begun in 1959 [Burkitt 1965]. It was a large system with provision for 120 control loops and 256 measurements, of which 98 and 224 respectively were used on the Fleetwood system. In 1961 the Monsanto Company also began a DDC project for a plant in Texas City and a *hierarchical* control scheme for the petrochemical complex at Chocolate Bayou.

The Ferranti Argus represented a change in computer hardware design in that the control program was held in a ferrite core store rather than on a rotating drum store as used by the RW-300 computer. The program was held in a programmable read-only memory; it was loaded by physically inserting pegs into a plug board, each peg representing one bit in the memory word. Although laborious to set up initially, the system proved to be very reliable in that destruction of the memory contents could only be brought about by the physical dislodgment of the pegs. In addition, security was enhanced by using special power supplies and switch-over mechanisms to protect information held in the main core store. This information was classified as follows:

1. *Set points* Loss most undesirable;
2. *Valve demand* Presence after controlled stoppage allows computer to gain control of plant immediately and without disturbance: *bumpless transfer*;
3. *Memory calculations* Loss is tolerable, soon will be updated and only slight disturbance to plant; and

4. *Future development calculation* Extension to allow for optimization may require information to be maintained for long periods of time.

In addition to improved reliability the Argus system provided more rapid memory access than the drum stores of the RW-300 and similiar machines and as such represented the beginning of the second phase of application of computers to real-time control.

The computers used in the early 1960s combined magnetic core memories and drum stores, the drums eventually giving way to hard disk drives. They included the General Electric 4000 series, IBM 1800, CDC 1700, Foxboro FOX 1 and 1A, the SDS and Xerox SIGMA series, Ferranti Argus series and Elliot Automation 900 series. The attempt to resolve some of the problems of the early machines led to an increase in the cost of systems: the increase was such that frequently their use could be justified only if both DDC and supervisory control were performed by the one computer. A consequence of this was the generation of further problems particularly in the development of the software. The programs for the early computers had been written by specialist programmers using machine code; and this was manageable because the tasks were clearly defined and the quantity of code relatively small. In combining DDC and supervisory control, not only had the quantity of code for a given application increased, but the complexity of the programming also increased in that the two tasks had very different time-scales; and the DDC control programs had to be able to interrupt the supervisory control programs. The increase in the size of the programs meant that not all the code could be stored in core memory: provision had to be made for the swapping of code between the drum memory and core.

The solution appeared to lie in the development of general purpose operating systems and high level languages. In the late 1960s real-time operating systems were developed and various PROCESS FORTRAN compilers made their appearance. The problems and the costs involved in attempting to do everything in one computer led users to retreat to smaller systems for which the newly developing minicomputer (DEC PDP-8 PDP-11, Data General Nova, Honeywell 316, etc.) were to prove ideally suited. The cost of the minicomputer was small enough to avoid the need to load a large number of tasks onto one machine; indeed by 1970 it was becoming possible to consider having two computers on the system, one simply acting as a stand-by in the event of failure.

The advent of the microprocessor in 1974 led to a further reappraisal of approach and the development of *distributed systems*. These developments are considered in more detail in Chapter 2.

1.2 ELEMENTS OF A COMPUTER CONTROL SYSTEM

As an example we shall consider a simple plant, a 'hot-air blower' as shown in Figure 1.2. A centrifugal fan blows air over a heating element and into a tube. A thermistor bead is placed at the outlet end of the tube and forms one arm of a bridge circuit. The

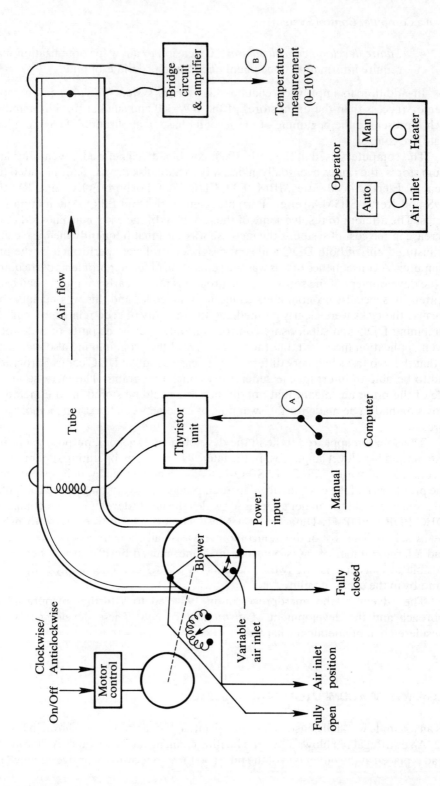

Fig 1.2 A simple plant: a hot-air blower.

amplified output of the bridge circuit is available at *B* and provides a voltage, in the range 0 to 10 volts, proportional to temperature. The current supplied to the heating element can be varied by supplying a dc voltage in the range 0 to 10 volts to point *A*.

The position of the air-inlet cover to the fan is adjusted by means of a reversible motor. The motor operates at constant speed and is turned on or off by a logic signal applied to its controller; a second logic signal determines the direction of rotation. A potentiometer wiper is attached to the air-inlet cover and the voltage output is proportional to the position of the cover. Microswitches are used to detect when the cover is fully open and fully closed.

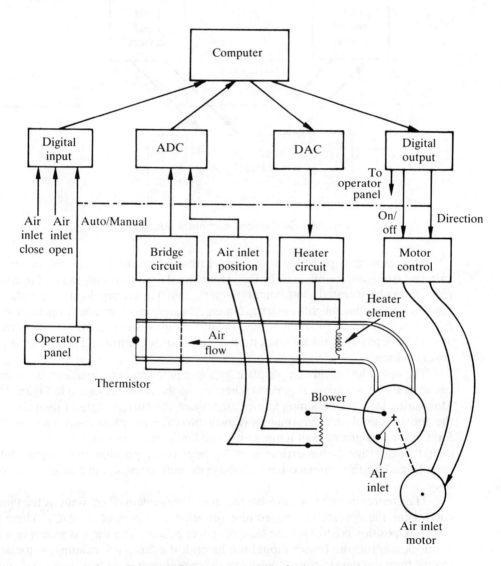

Fig. 1.3 Computer control of a hot-air blower.

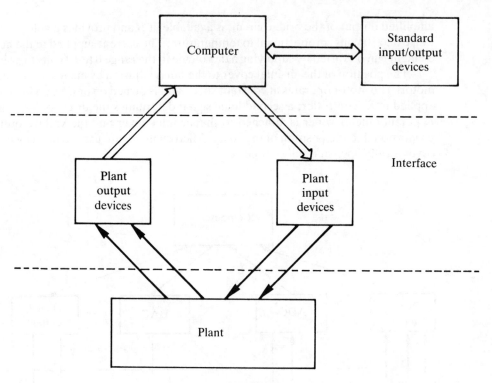

Fig. 1.4 Generalized computer control system.

The operator is provided with a panel from which the control system can be switched from auto to manual. In the manual mode the heat output and fan cover position can be adjusted using potentiometers. Switches are provided to operate the fan and heater. Panel lights indicate 'fan on', 'heater on', 'cover fully open', 'cover fully closed' and 'auto/manual' status. In auto status, slider potentiometer controls are available to adjust the set point for the temperature control and to adjust the fan cover position.

The operation of this simple plant by a computer requires *monitoring, control calculation* and *actuation*. A general schematic of the system is shown in Figure 1.3. Monitoring involves obtaining information about the current state of the plant. In the above example the information is available from the plant instruments in the form of analog signals for air temperature and fan-inlet cover position; in the form of digital (logic) signals for extremes of fan-inlet cover position (fully open, fully closed); and for the various other status signals: auto/manual, fan motor on, heater on.

The control calculations involve the digital equivalent of continuous feedback control for the control of temperature (direct digital control – DDC). There is feedback-position control for the fan-inlet cover position and there is sequence and interlock control: the heater should not be on if the fan is not running; automatic change from tracking to controlling when the operator changes from manual to auto

Elements of a Computer Control System

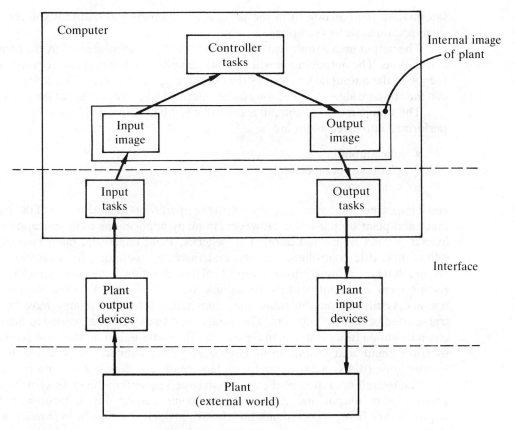

Fig. 1.5 Generalized computer control system showing hardware and software interface.

(these actions may involve parallel logic operations, time-sequential control and timing of operations).

The actuation requires the provision of a voltage proportional to the demanded heat output for the heater control; logic signals indicating on/off and direction for the fan-inlet cover and logic signals for the operator display.

The monitoring and actuation tasks thus involve a range of interface devices including analog-to-digital converters, digital-to-analog converters, digital input and output lines and pulse generators. Details of the different types of interfaces are given in Chapter 3. For the present we will represent them simply as input and output devices as shown in Figure 1.4. Each of the various types of device will require software to operate it; again we will represent this software as input and output tasks. This generalized picture of a computer control system is shown in Figure 1.5.

The input devices plus the input software provide the information to create an 'input image' of the plant. The input image is a snapshot of the status of the plant and this snapshot is renewed at specified intervals. (All of it may be renewed at the same time or some parts may be renewed less frequently than others.) Where the external information is in analog form the process of obtaining the snapshot will involve quantization as well as sampling.

The output image represents the current set of outputs generated by the control calculations. The output image will be updated periodically by the control tasks. It is the job of the output task to convey the output image to the plant. The control tasks can thus be considered as operating on an internal image (or model) of the plant.

The simple model of computer control described above divides the tasks to be performed into three major areas:

- plant input tasks,
- plant output tasks, and
- control tasks;

and it is assumed that communication with the operator is treated as part of the plant input and plant output tasks. However, in many applications communication will extend beyond simple indicators and switches. Plant engineers, plant managers, pilots, air-traffic controllers, drivers, and machine operators, for example, will require detailed information on all aspects of the operation of the plant, aircraft, car, radar system, etc. Control of the system may be shared between several computers, not necessarily all on the same site, and hence information may have to be transmitted between computers. The model must therefore be extended to include communication tasks as shown in Figure 1.6. The communication tasks are assumed to cover input and output from keyboards/VDUs, remote transmission links, backing store (disks, etc.), output to printers, chart recorders, graph plotters, etc.

The overall operation of the system may be sequential: the tasks plant input, control, plant output and communication being carried out in turn, with the sequence then being repeated indefinitely; or the operation may be in parallel, with the various tasks being carried out concurrently. In the latter case problems of synchronizing the various tasks occur. Both options are explored in detail in Chapter 5.

1.3 CLASSIFICATION OF REAL-TIME SYSTEMS

The plant input, plant output, and communication tasks shown in Figure 1.6 have one feature in common – they are connected by physical devices to processes which are external to the computer. These external processes all operate in their own time-scales and the computer is said to operate in real time if actions carried out in the computer relate to the time-scales of the external processes.

The relationship may be defined in terms of the passage of time, or the actual time of day, in which case the system is said to be 'clock-based'. Or it may be defined in terms of events, e.g. the closure of a switch, in which case it is said to be sensor- or event-based. There is a third category, interactive, in which the relationship between the actions in the computer and the system is much more loosely defined. Typically, the requirement is that a set of operations in the computer should be completed within a predetermined time. The majority of communication tasks fall into this category.

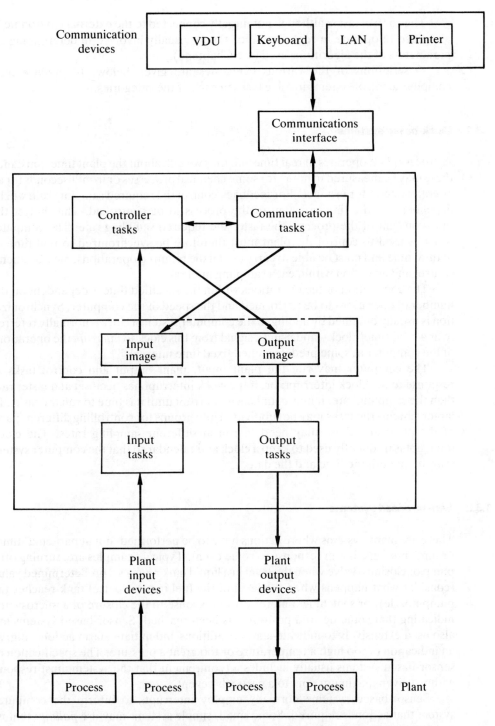

Fig. 1.6 Computer control system showing communication tasks.

The control tasks, although not directly connected to the external environment, also need to operate in real time since time is usually involved in determining the parameters of the algorithms used (see Chapter 4).

Descriptions of the various categories are given below. In practice most computer control systems involve tasks in each of the categories.

1.3.1 Clock-based systems

A process plant operates in real time and thus we talk about the plant time-constants; these may be measured in hours for some chemical processes or in milliseconds for an aircraft system, for instance. For feedback control the required sampling rate will be dependent on the time-constant of the process to be controlled. The shorter the time-constant of the process, the faster the required sampling rate. The computer which is used to control the plant must therefore be synchronized to real time or natural time and must be able to carry out all the required operations, measurement, control and actuation within each sampling interval.

The completion of the operations within the specified time is dependent on the number of operations to be performed and the speed of the computer. Synchronization is usually obtained by adding to the computer system a clock – normally referred to as a 'real-time' clock – and using a signal from this clock to *interrupt* the operations of the computer at some predetermined fixed time interval.

The computer may run the plant input, plant output and control tasks in response to the clock interrupt: or, if the clock interrupt has been set at a faster rate than the sampling rate, it may count each interrupt until it is time to run the tasks. In larger systems the tasks may be subdivided into groups for controlling different parts of the plant and these may need to run at different sampling rates. The clock interrupt is frequently used to keep a clock and calendar so that the computer system is aware of both the time and the date.

1.3.2 Sensor-based systems

There are many systems where actions have to be performed, not at particular times or time intervals, but in response to some event. Typical examples are: turning off a pump or closing a valve when the level in a liquid tank reaches a predetermined value (consider what happens when the level of the fuel in a car petrol tank reaches the pump nozzle); or switching a motor off in response to the closure of a microswitch indicating that some desired position has been reached. Sensor-based systems are also used extensively to indicate alarm conditions and initiate alarm actions, e.g. as an indication of too high a temperature or too great a pressure. The specification of sensor-based systems usually includes a requirement that the system must respond within a given maximum time to a particular event.

Sensor-based systems normally employ interrupts to inform the computer system that action is required. Some small, simple systems may use *polling*; that is, the computer periodically asks (polls) the various sensors to see if action is required.

1.3.3 Interactive systems

Interactive systems probably represent the largest class of real-time systems and cover such systems as automatic bank tellers; reservation systems for hotels, airlines and car rental companies; computerized tills, etc. The real-time requirement is usually expressed in terms of the average response time not exceeding a specified value. For example, an automatic bank teller system might require an average response time not exceeding 20 seconds. Although this type of system superficially seems similar to the sensor-based system – that is, it apparently responds to a signal from the plant (in this case usually a person) – it is different in that it responds at a time determined by the internal state of the computer. An automatic bank teller is not interested in the fact that you are about to miss a train or that it is raining hard and you are getting wet; this is irrelevant to its response.

Many interactive systems give the impression that they are clock-based in that they are capable of displaying the data and time; they do indeed have a real-time clock which enables them to keep track of time. The test as to whether or not they are clock-based as described above is to ask, 'Can they be tightly synchronized to an external process?'

1.4 REAL-TIME SYSTEMS – A DEFINITION

In the rest of this book the term 'real time' will be used to refer to systems in which:

(1) the order of computation is determined by the passage of time or by events external to the computer; and
(2) the results of the particular calculation may depend upon the value of the variable 'time' at the instance of execution of the calculations *or* the time taken to execute the computation.

Following the classification arrangement of Civera et al. [1983], real-time systems will be divided into two categories based on a definition of the system functioning correctly:

Type 1 The system must have a mean execution time measured over a defined time interval which is lower than a specified maximum.
Type 2 The computation must be completed within a specified maximum time on each and every occasion.

The second category is obviously a much more severe constraint on the performance of the system than the first. It is typical of the so-called 'embedded systems'; that is, systems in which the computer is (or computers are) an integral part of some machine. Such systems present a difficult challenge both to hardware and to software designers.

An example of a type 1 real-time system is an automatic bank teller. It is real time since (1) it is driven by external events: placing your card into the machine initiates the transaction; (2) the computation performed depends on the variable time: typically you are limited to a fixed maximum withdrawal per day or per week; and (3) the constraint on the system is based on mean response time not a hard constraint for every transaction (compare the response time obtained between 12 and 2 p.m. on a Friday with that at 10 a.m. on a Sunday). There are also other constraints on the system; for example, it will be expected to debit the correct account.

It is necessary to make a careful distinction when using 'external event driven' as a classifying condition. In one sense, all interactive programs are driven by external events: the distinction which has to be made is that in a real-time program the external event, although limited to a certain class, cannot be predicted from the current state of the program. For example, the automatic bank teller software cannot know which customer and which unit will next demand attention. In an interactive program the software knows that the response will come from a particular terminal and that it will belong to a certain class of responses (any response which does not belong to the class is an error). The behavior of the automatic teller system apparently becomes interactive once the customer's card has been accepted and read, since it then presents a series of questions to be answered. It is, however, still not a purely interactive system in that the action taken in response to the answers depends on time.

A typical example of a type 2 real-time control system is the temperature control loop of the hot-air blower system described above. In control terms the temperature loop is a sampled data system which can be represented in block diagram form as shown in Figure 1.7. Design of a suitable control algorithm for this system involves the choice of the sampling interval T_s. This sampling interval must form part of the specification of the software for implementing the control algorithm. The specification of the hot-air blower might say that the sampling rate must be 100 Hz (i.e., a sample interval of 10 ms), not that the average response time must be 10 ms. This would then require that every 10 ms the input value must be read and the controller value calculated and then sent to the output, and that this process must not be delayed by any other activities.

In actual practice the above categories are only guides; for example, in the hot-air blower an occasional missed sample would not seriously affect the performance, neither would a variation in sampling interval such that inequality $9.95 \leq T_s \leq 10.05$ ms with a mean of $T_s = 10$ ms was satisfied. The system would not be satisfactory, however, if the sampling interval T_s was in the range $1 < T_s < 1000$ ms with a mean of 10 ms over a 24-hour period. Similarly, the automatic bank teller would not be satisfactory if at busy times customers had to wait ten minutes, even if it achieved a mean response measured over a 24-hour period of 20 seconds. A typical specification might be a mean response time over a 24-hour period of 15 seconds, with 95% of requests being satisfied within 30 seconds and no response time greater than 60 seconds.

A second time constraint, which is concerned with response to alarm conditions, is often associated with a type 2 system. This is normally a hard constraint of the form that the computer must respond within some specified time. The hot-air blower for example may be being used to dry a component which will be damaged if exposed to temperatures greater than 50°C for more than 10 seconds. Allowing for the time

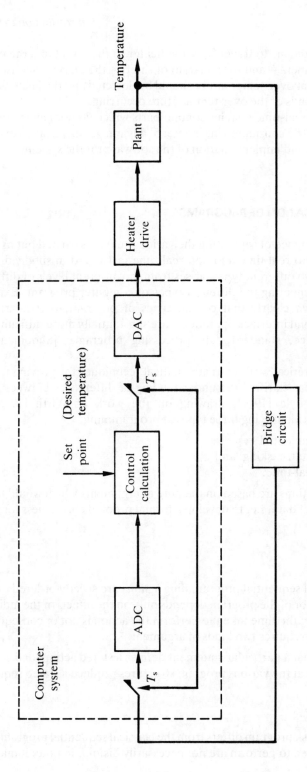

Fig. 1.7 Block diagram of computer control system.

taken for the air to travel from the heater to the outlet and the cooling time of the heater element – and for a margin of safety – the alarm response requirement may be, say, that over-temperature should be detected and the heater switched off within seven seconds of the over-temperature occurring.

Most real-time systems contain parts which do not have to operate in real time and, as will be considered in Chapter 5, advantage can be taken of this fact to simplify the design and implementation of the software for the system.

1.5 CLASSIFICATION OF PROGRAMS

The importance of separating the various activities carried out by computer control systems into real-time and non-real-time tasks, and in subdividing real-time tasks into the two different types, arises from the different levels of difficulty of designing and implementing the different types of computer program. Experimental studies have shown clearly that certain types of program, particularly those involving real-time and interface operations, are substantially more difficult to construct than, for instance, standard data processing programs: [Shooman 1983, Pressman 1982].

Theoretical work on mathematical techniques for proving the correctness of a program has clarified the understanding of differences between different types of program. Pyle [1979], drawing on the work of Wirth [1975], has presented definitions identifying three types of programming:

- sequential;
- multi-tasking; and
- real-time.

The definitions are based on the kind of arguments which would have to be made in order to validate, i.e. to develop a formal proof of correctness for, programs of each type.

Sequential

In classical sequential programming *actions* are strictly ordered as a time sequence: the behavior of the program depends only on the effects of the individual *actions* and their *order*; the time taken to perform the action is not of consequence. Validation, therefore, requires two kinds of argument:

1. that a particular statement defines a stated action; and
2. that the various program structures produce a stated sequence of events.

Multi-tasking

A multi-task program differs from the classical sequential program in that the actions it is required to perform are not necessarily disjoint in time: it may be necessary for several actions to be performed in parallel. We should note that the sequential relationships between the actions may still be important. Such a program may be built from a number of parts (processes or tasks are the names used for the parts), which are themselves partly sequential, but which are executed concurrently and which communicate through shared variables and synchronization signals.

Validation requires the application of arguments for sequential programs with some additions. The tasks (processes) can be validated separately only if the constituent variables of each task (process) are distinct. If the variables are shared, the potential concurrency makes the effect of the program unpredictable (and hence not capable of validation) unless there is some further rule that governs the sequencing of the several actions of the tasks (processes). It should be noted that the use of a synchronizing procedure means that the time taken for each individual action is not relevant to the validation procedure. The task can proceed at any speed: the validity depends on the actions of the synchronizing procedure. (Synchronization techniques are discussed extensively in Chapter 7.)

Real time

A real-time program differs from the previous types in that, in addition to its actions not necessarily being disjoint in time, the sequence of some of its actions is not determined by the designer but by the environment (that is, by events occurring in the outside world; events which occur in real time and without reference to the internal operations of the computer). Such events cannot be made to conform to the inter-task synchronization rules. A real-time program can still be divided into a number of tasks, but communication between the tasks cannot necessarily wait for a synchronization signal: the environment task cannot be delayed. (It should be noted that in process control applications the main environment task is usually that of keeping real time; i.e., a real-time clock task. It is this task which provides the timing for the scanning tasks which gather information from the outside world about the process.) In real-time programs, in contrast to the two previous types of program, the *actual time taken* by an action is an essential factor in the process of validation.

Consideration of the types of reasoning necessary for the validation of programs is important, not because we, as engineers, are seeking a method of formal proof, but because we are seeking to understand the factors which need to be considered when designing real-time software. It has been found by experience that the design of real-time software is significantly more difficult than the design of sequential software. The problems of real-time software design have not been helped by the fact that the early high-level languages were sequential in nature and they did not allow direct access to many of the detailed features of the computer hardware. As a consequence, real-time features had to be built into the operating system which was written in the assembly language of the machine by teams of specialist programmers. The cost of producing such operating systems was high and they had therefore to be general purpose so that they could be used in a wide range of applications in order to reduce the unit cost of producing them. These operating systems could be 'tailored'; i.e., they could be reassembled to exclude or include certain features, to change the number of tasks which could be handled, or to change the number of input/output devices and types of device for example. Such changes could usually only be made by the supplier.

1.6 SUMMARY

In this chapter a brief history of computer control has been given; information on further development and the reasons for particular computer structures is given in Chapter 2. The simple example of a hot-air blower has been used to illustrate the breadth of knowledge required to design and implement a computer control system. The engineer is required to have knowledge of:

- the plant;
- transducers;
- actuators;
- computer hardware;
- interface techniques;
- communication systems;
- software design;
- programming languages;
- control algorithm design; and
- signal processing.

In this book we do not attempt to cover all these areas in detail, but instead concentrate on software design and the implementation of software.

As should have already been apparent, there will be an emphasis on dividing the operations to be performed into separate tasks. The detail of operations within the tasks will be hidden within the software for that task. This is a process known as 'abstraction'. The organization of the tasks to form an overall system can then be carried out without the distraction the detail of their operations would otherwise provide.

Essential information on the hardware of computers and interface devices is given in Chapter 3 while the problems of implementing control algorithms are dealt with in Chapter 4.

EXERCISES

1.1 You have been asked to design a computer-based system to control all the operations of a retail petrol (gasoline) station (control of pumps, cash receipts, sales figures, deliveries, etc.). What types of real-time system would you expect to use?

1.2 An automatic bank teller works by polling each teller in turn. Some tellers are located outside buildings and others inside. How could the polling system be organized to ensure that the waiting time at the outside locations was less than at the inside locations?

1.3 Would you classify any of the following systems as real-time? In each case give reasons for your answer and classify the real-times systems as type 1 or type 2.
(a) A simulation program run by an engineer on a personal computer.
(b) An airline seat-reservation system with on-line terminals.
(c) A microprocessor-based automobile ignition and fuel injection system.
(d) A computer system used to obtain and record measurements of force and strain from a tensile strength testing machine.
(e) An aircraft autopilot.

1.4 For (a) the petrol (gas) station system in **1.1** and (b) the automatic bank teller in **1.2**, list the activities which have to be performed and estimate the time requirements of each activity.

1.5 For the hot-air blower described in section 1.2 estimate the precision required for the analog-to-digital and digital-to-analog converters.

REFERENCES AND BIBLIOGRAPHY

ANON (1959), 'Computing control – a commercial reality', *Control Engineering,* **9**(5) p. 40
AURICOSTE, J.G. (1963), 'Applications of digital computers to process control', in Coales, J. (ed.) *Automation in the Chemical, Oil and Metallurgical Industries*, Butterworths
BENNETT, S. and LINKENS, D.A. (eds.) (1982), *Computer Control of Industrial Processes*, Peter Peregrinus, Stevenage
BENNETT, S., and LINKENS, D.A. (eds.) (1984), *Real-time Computer Control*, Peter Peregrinus, Stevenage
BIBBERO, R.J. (1977), *Microprocessors in Instruments and Control*, Wiley
BROWN, G.S. and CAMPBELL, D.P. (1950), 'Instrument engineering: its growth and promise in process-control problems', *Mechanical Engineering* **72**(2), p. 124
BURKITT, J.K. (1965), 'Reliability performance of an on-line digital computer when controlling a plant without the use of conventional controllers', *Automatic Control in the Chemical Process and Allied Industries*, Society of Chemical Industry, pp. 125–40
CASSELL, D.A. (1983), *Microcomputers and Modern Control Engineering*, Prentice Hall
CHARD, R.A. (1983), *Software Concepts in Process Control*, NCC Publications
CIVERA, P., DEL CORSO, D. and GREGORETTI, F. (1983), 'Microcomputer systems in real-time applications', in Tzafestas, S.G. (ed.), *Microprocessors in Signal Processing, Measurement and Control*, Reidel
DESHPANDE, P.B. and ASH, R.H. (1983). *Elements of Computer Process Control*, Prentice Hall
FREEDMAN, A.L. and LEES, R.A. (1977), *Real-time Computer Systems*, Edward Arnold
HOLLAND, R.C. (1983), *Microcomputers for Process Control*, Pergamon
JOHNSON, C.D. (1984), *Microprocessor-based Process Control*, Prentice-Hall
JOVIC, F. (1986), *Process Control Systems: Principles of Design and Operation*, Kogan Page
LOWE, E.I. and HIDDEN, A.E. (1971), *Computer Control in Process Industries*, Peter Peregrinus, Stevenage
MELLICHAMP, D.A. (ed.) (1983), *Real-time Computing with Applications to Data Acquisition and Control*. Van Nostrand
PRESSMAN, R.S. (1982), *Software Engineering: a Practitioner's Approach*, McGraw-Hill

PYLE, I.C. (1979), 'Methods for the design of control software', in *Software for Computer Control. Proc. Second IFAC/IFIP Symposium on Software for Computer Control*, Prague 1979, Pergamon, Oxford

SAVAS, E.S. (1965), *Computer Control of Industrial Processes*, McGraw-Hill.

SHOOMAN, M.L. (1983), *Software Engineering*, McGraw-Hill

STIRE, T.G. (ed.) (1983), *Process Control Computer Systems: Guide for Managers*, Ann Arbor.

TZAFESTAS, S.G. (ed.) (1983), *Microprocessors in Signal Processing, Measurement and Control*, Reidel, Dordrecht

WILLIAMS, T.J. (1977), 'Two decades of change', in *Control Engineering*, **24**(9) pp. 71–6

2

Concepts of Computer Control

2.1 INTRODUCTION

Industrial and laboratory processes (and most systems in the widest sense) which use a computer or computers in their operation can be classified under one or more of the following categories of operation:

- batch;
- continuous; and
- laboratory (or test).

The categories are not mutually exclusive – a particular process may involve activities which fall into more than one of the above categories; they are, however, useful for describing the general character of a particular process.

The term 'batch' is used to describe processes in which a sequence of operations is carried out to produce a quantity of a product – the batch – and in which the sequence is then repeated to produce further batches. The specification of the product or the exact composition may be changed between the different runs. A typical example of batch production is rolling of sheet steel – an ingot is passed through the rolling mill to produce a particular gauge of steel. The next ingot may either be of a different composition or be rolled to a different thickness and hence will require different settings of the rolling mill. A characteristic of batch processing is 'set-up' time (or 'down time' or 'changeover' time), i.e. the time taken to prepare the equipment for the next production batch. This is 'wasted time' in that no output is being produced; the ratio between 'operation time' (the time during which the product is being produced) and 'set-up' time is important in determining a suitable batch size. In mechanical production the advent of the NC (numerically controlled) machine tool which can be set up in a much shorter time than the earlier 'automatic' has led to a reduction in the size of batch considered to be economic. Aircraft control sysems fall into this category as each separate flight can be thought of as equivalent to a batch.

The term 'continuous' is used for systems in which production is maintained for long periods of time without interruption, typically over several months or even years. An example of a continuous system is the catalytic cracking of oil in which the crude oil enters at one end and the various products – fractionates – are removed as the process continues. The ratio of the different fractions can be changed but this is done without halting the process. 'Continuous' systems may produce 'batches' in that the product composition may be changed from time to time, but they are still classified as continuous since the change in composition is made without halting the production process. A problem which occurs in continuous processes is that, during changeover from one specification to the next, the output of the plant is often not within the product tolerance and must be scrapped, hence it is financially important that the change be made as quickly and smoothly as possible.

There is a trend to convert processes to continuous operation – or, if the whole process cannot be converted, to convert part of the process. For example, in the baking industry bread dough is produced in batches, but continuous ovens are frequently used to bake it whereby the loaves are placed on a conveyor which moves slowly through the oven. An important problem in mixed mode systems, i.e. systems in which batches are produced on a continuous basis, is the tracking of material through the process; it is obviously necessary to be able to identify a particular batch at all times.

Laboratory-based systems are frequently of the 'operator-initiated' type in that the computer is used to control some complex experimental test or some complex equipment used for routine testing. A typical example is the control and analysis of data from a vapour phase chromatograph. Another example is the testing of an audiometer, a device used to test hearing. The audiometer has to produce sound levels at different frequencies; it is complex in that the actual level produced is a function of frequency since the sensitivity of the human ear varies with frequency. Each audiometer has to be tested against a sound-level meter and a test certificate produced. This is done by using a sound-level meter connected to a computer and using the output from the computer to drive the audiometer through its frequency range. The results printed out from the test computer provide the test certificate.

As with attempts to classify systems as batch or continuous so it can be difficult at times to classify systems solely as laboratory. The production of steel using the electric arc furnace involves complex calculations to determine the appropriate mix of scrap, raw materials and alloying additives. As the melt progresses samples of the steel are taken and analyzed using a spectrometer. Typically this instrument is connected to a computer which analyzes the results and calculates the necessary adjustment to the additives. The computer used may well be the computer which is controlling the arc furnace itself.

In all these systems there will be several different computer activities being carried out:

- sequence control;
- loop control (DDC);
- supervisory control;
- data acquisition;
- data analysis; and
- human interfacing (MMI).

The overall objectives of the control of these processes can be summarized as:

- safety;
- product specification;

- environment;
- ease of operation; and
- economics.

The aim of introducing computer control is to seek improvements in all or some of these areas.

2.2 SEQUENCE CONTROL

Although sequence control will occur in some part of most systems it often predominates in batch systems and hence a batch system is used to illustrate it.

Batch systems are widely used in the food-processing and chemical industries where the operations carried out frequently involve mixing raw materials, carrying out some process and then discharging the product. A typical reactor vessel for this purpose is shown in Figure 2.1. A chemical is produced by the reaction of two other chemicals at a specified temperature. The chemicals are mixed together in a sealed vessel (the reactor) and the temperature of the reaction is controlled by feeding hot or cold water through the water jacket which surrounds the vessel. The water flow is controlled by adjusting valves C and D. The flow of material into and out of the vessel is regulated by the valves A, B and E. The temperature of the contents of the vessel and the pressure in the vessel are monitored.

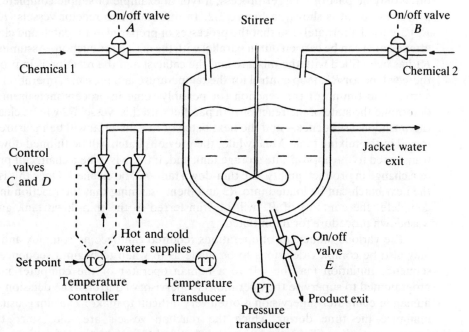

Fig. 2.1 A simple chemical reactor vessel (redrawn from Bennett and Linkens, *Real time Computer Control*, 1984, Peter Peregrinus).

The procedure for the operation of the system may be as follows:

1. Open valve A to charge the vessel with chemical 1.
2. Check the level of the chemical in the vessel (by monitoring the pressure in the vessel); when the correct amount of chemical has been admitted, close valve A.
3. Start the stirrer used to mix the chemical together.
4. Repeat stages 1 and 2 with valve B in order to admit the second chemical.
5. Switch on the three-term controller and supply a set point so that the chemical mix is heated up to the required reaction temperature.
6. Monitor the reaction temperature: when it reaches the set point, start a timer to time the duration of the reaction.
7. When the timer indicates that the reaction is complete, switch off the controller and open valve C to cool down the reactor contents. Switch off the stirrer.
8. Monitor the temperature; when the contents have cooled, open valve E to remove the product from the reactor.

When implemented by computer all of the above actions and timings would be based upon software. For a large chemical plant such sequences could become very lengthy and intricate and for plant efficiency a number of sequences may be taking place in parallel.

The processes carried out in the single reactor vessel shown in Figure 2.1 would often only be part of a larger process; a typical example of simple complete batch processing plant is shown in Figure 2.2. In this plant two reactor vessels ($R1$ and $R2$) are used alternately, so that the processes of preparing for a batch and cleaning after a batch can be carried out in parallel with the actual production. Assuming that $R1$ has been filled with the mixture and the catalyst and the reaction is in progress, there will be for $R1$: loop control for the temperature and pressure; operation of the stirrer; and timing of the reaction (or possibly some in-process measurement to determine the state of the reaction). In parallel with this, vessel $R2$ will be cleaned – the washdown sequence – and the next batch of raw material will be measured and mixed in the mixing tank. Meanwhile, the previous batch will be thinned down and transferred to the appropriate storage tank and, if there is to be a change of product or a change in product quality, the thin-down tank will be cleaned. Once this is done the next batch can be loaded into $R2$ and then, assuming that the reaction in $R1$ is complete, the contents of $R1$ will be transferred to the thin-down tank and the washdown procedure for $R1$ initiated.

The various sequences of operations required can become complex and there may also be complex decisions to be made as to when to begin a sequence. The sequence initiation may be left to a human operator or the computer may be programmed to supervise the operations (supervisory control). The decision to use human or computer supervision is often very difficult to make. The aim is usually to minimize the time during which the reaction vessels are idle, since this is unproductive time. The calculations needed and the evaluation of options can be complex, particularly if, for example, reaction times vary with product mix and

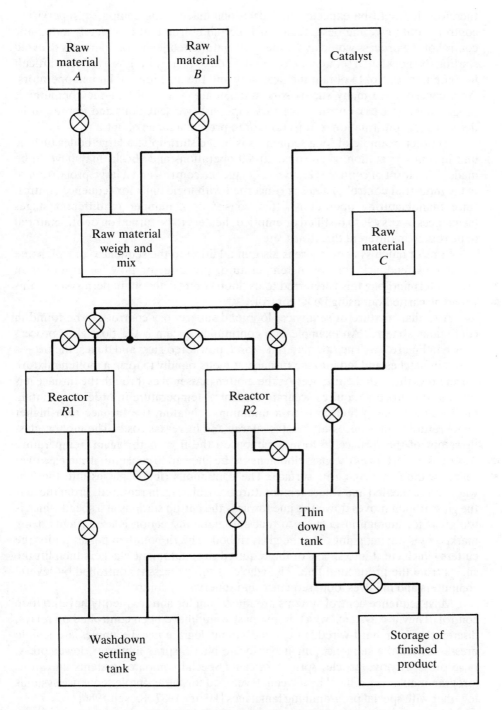

Fig. 2.2 Typical chemical batch process.

therefore it would be expected that decisions made using computer 'supervisory control' would give the best results. It is certainly true that completely automatic control of the process would avoid the quite natural tendency for operators to avoid starting the next batch sequence close to the end of their shift; however, it is difficult in computer control to obtain the flexibility and the ingenuity of human operators. As a consequence many supervisory systems are mixed in that the computer is programmed to carry out the necessary supervisory calculations and to present its decisions for confirmation or rejection, or to present a range of options.

Another example of long sequences is in the start up of a large boiler turbine unit in a power station when some 20 000 operations and checks may have to be made. This form of control is known as 'sequence control' and a large proportion of many industrial control systems is concerned with logic units for sequence control, since manufacturing operations often consist of a number of different stages (sequences) – vessels to be filled or emptied, heaters to be turned on or off, material to be routed in different directions, etc.

In most batch systems there is also, in addition to the sequence control, some continuous feedback control of temperature, pressure, or flow for example. In process terminology this is referred to as 'loop control' and in modern systems this would be carried out using DDC control.

A similar mixture of sequence, loop and supervisory control can be found in continuous systems. An example of a continuous system is the float glass process shown in Figure 2.3. The raw material – sand, powdered glass and fluxes, the frit – is mixed in batches and fed into the furnace; it melts rapidly to form a molten mixture which flows through the furnace. As the molten glass moves through the furnace it is refined; the process requires accurate control of temperature in order to maintain quality and to keep fuel costs to a minimum – heating the furnace to a higher temperature than is necessary wastes energy and increases costs. The molten glass flows out of the furnace to form a ribbon on the float bath; again, temperature control is important as the glass ribbon has to be allowed to cool until it can pass over rollers without damaging its surface. The continuous ribbon passes into the lehr where it is annealed and where temperature control is again required. From the lehr the glass ribbon moves down the line towards the cut-up stations at a speed which is too great for manual inspection so that automatic inspection is used, faults being marked by spraying paint onto the glass ribbon. The ribbon then passes under the cutters which cut it into sheets of the required size; automatic stackers then lift the sheets from the production line. The whole of this process is controlled by several computers and involves loop, sequence and supervisory control.

Most sequence control systems are much simpler and frequently have no loop control. They are systems which in the past would have been controlled by relays, discrete logic or hard-wired, integrated circuit logic units. Examples are simple presses where the sequence might be: locate blank, spray lubricant, lower press, raise press, remove article, spray lubricant. Special computer systems known as 'programmable controllers' have been developed for these simple sequence systems together with special programming languages [Henry 1987, Kissell 1986].

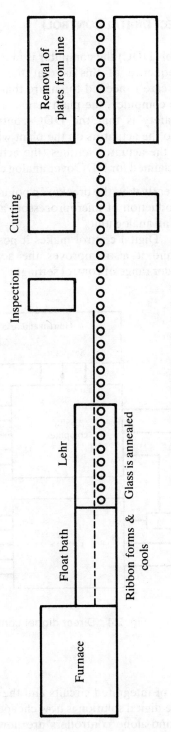

Fig. 2.3 Schematic of float glass process.

2.3 LOOP CONTROL (DIRECT DIGITAL CONTROL)

In direct digital control (DDC) the computer is in the feedback loop as is shown in Figure 2.4. A consequence of this is that the computer becomes a critical component and great care is needed to ensure that, in the event of the failure or malfunctioning of the computer, the plant remains in a safe condition. The usual means of ensuring safety is that the DDC controller is restricted to making incremental changes to the actuators on the plant while a limit is usually placed on the rate of change of the actuator settings (the actuators are labeled A in Figure 2.4). The advantages claimed for DDC over analog control are:

1. Cost In the early days the breakeven point was between 50 and 100 loops; with the introduction of microprocessors a single loop DDC unit can be cheaper than an analog unit.
2. Performance Digital control makes it possible to use improved control algorithms, and it also improves the accuracy of the controller and provides a wider range of control settings.

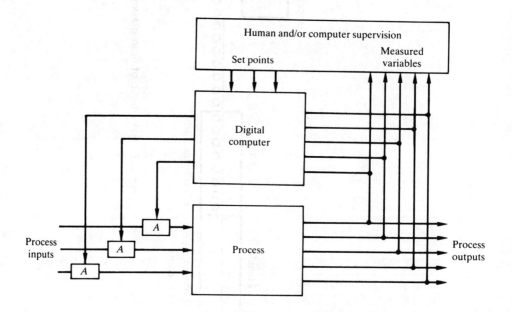

Fig. 2.4 Direct digital control.

The development of integrated circuits and the microprocessor have ensured that in terms of cost the digital solution is now cheaper than the analog. Single loop controllers used as stand-alone controllers are now based on the use of digital techniques and contain one or two microprocessor chips which are used to implement DDC control algorithms. Many are designed such that they can be used directly to replace existing analog control units.

The adoption of improved control algorithms has, however, been slow. Many computer control implementations have simply taken over the well-established analog three-term control algorithm (proportional + integral + derivative – PID). This has the general form

$$m = K_p (e + 1/T_i\, e.dt + T_d.\, de/dt) \qquad (2.1)$$

where $e = r - y$ and y is the measured variable, r is reference value or set point, and e is error; K_p is the overall controller gain; T_i is the integral action time; and T_d is the derivative action time. This algorithm can be expressed in other forms. For example, derivative action is frequently not used, when de/dt is often replaced by dy/dt to avoid differentiating the set point.

The algorithm is normally implemented in DDC systems by using the difference equation equivalent to 2.1 above. If the sampling interval for computation is T seconds then the simple approximations

$$de/dt = (e_k - e_{k-1})/T \quad \text{and} \quad e.dt = e_k.\, T \quad k = 0, 1, 2 \ldots$$

can be used. The control equation then becomes

$$m_k = K_p \{e_k + T/T_i.S_k + T_d/T(e_k - e_{k-1})\}$$

with $S_k = S_{k-1} + e_k$ being the sum of the errors.

This is at first sight a simple algorithm to implement; however, there are many traps into which the unwary engineer can fall. One of the most dangerous is associated with the saturation of the actuation signal. The actuation signal has to lie between two limits U_{min} and U_{max} (constraints imposed by the physical properties of the actuator mechanism); if the signal exceeds these limits at either extreme, careful consideration has to be given to the integral sum S_k. For example, if the summation procedure is allowed to continue unchecked S can reach large values which could lead to a much degraded control system performance. Several methods have been developed for dealing with this problem and they will be dealt with later (see Chapter 4). Other problems of implementation concern word length and sampling rates. Again these will be considered in more detail later (see Chapter 4).

DDC control may be applied either to a single loop system implemented on a small microprocessor or to a large system involving several hundred loops. The loops may be cascaded, that is, with the output or actuation signal of one loop acting as the set point for another loop, signals may be added together (ratio loops) and conditional switches may be used to alter signal connections. A typical industrial system is shown in Figure 2.5. This is a steam boiler control system: the requirements are that the steam pressure is to be controlled by regulating the supply of fuel oil to the burner; however, in order to comply with the pollution regulations a particular mix of air and fuel is required. We are not concerned with how this is achieved but with the elements which are required to implement the chosen control system.

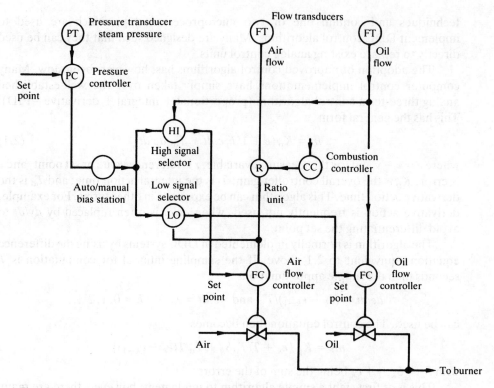

Fig. 2.5 A boiler control scheme (redrawn from Bennett and Linkens, *Real-time Computer Control*, 1984, Peter Peregrinus).

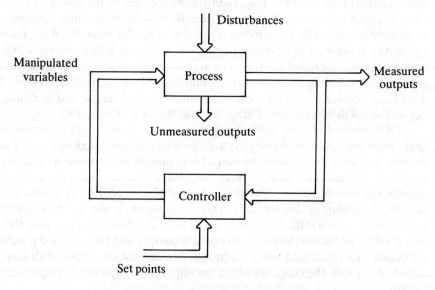

Fig. 2.6 General structure of a feedback control configuration.

Sequence Control

The steam pressure control system generates an actuation signal which is fed to an auto/manual bias station. If the station is switched to auto then the actuation signal is transmitted; if it is in manual mode a signal which has been entered manually (say from a keyboard) is transmitted. The signal from the bias station is connected to two units – a high signal selector and a low signal selector; each selector has two inputs and one output. The high selector transmits the higher of the two input signals, the low selector transmits the lower of the two inputs. The signal from the low selector provides the set point for the DDC loop controlling the oil flow, the signal from the high selector provides the set point for the air flow controller (two cascade loops). A ratio unit is installed in the air flow measurement line. A signal from the controller which monitors the combustion flames directly (using an optical pyrometer) is added to the air flow signal to provide the input to the air flow controller.

There is much more to computer control than simple DDC and the implementation of the control algorithms. The use of DDC is not limited to the use of the three-term control algorithm; algorithms developed using various digital control design techniques can be equally effective and a lot more flexible than the three-term controller. However, the art of tuning three-term controllers is well established and the technique gives a well behaved controller so that the introduction of new techniques is slow. In fact, because the three-term controller copes perfectly adequately with 90% of all control problems, it provides a strong deterrent to the adoption of new control system design techniques.

DDC is not necessarily limited to simple feedback control, as is shown in Figure 2.6; it is possible to use techniques such as inferential, feedforward and adaptive or self-tuning control. Inferential control, illustrated in Figure 2.7, is the term applied

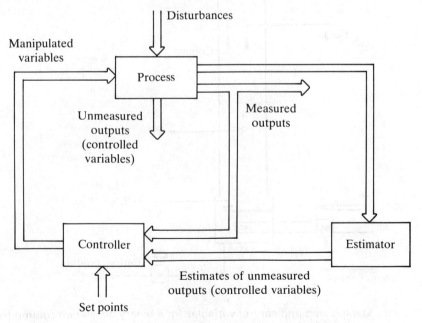

Fig. 2.7 General structure of inferential control configuration.

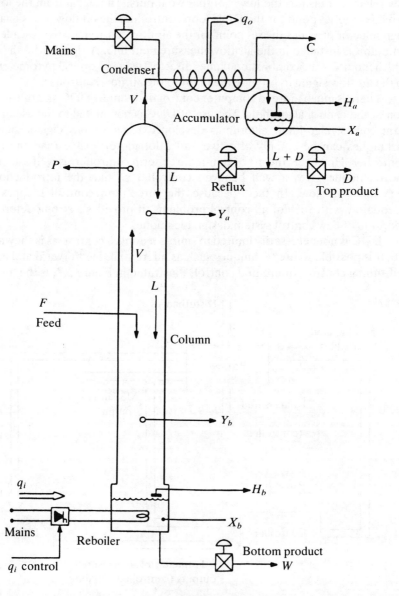

Fig. 2.8 Manipulated and control variables for a binary distillation column (redrawn from Bennett and Linkens, *Computer Control of Industrial Processes*, 1982, Peter Peregrinus).

to control where the variables on which the feedback control is to be based cannot be measured directly, but have to be 'inferred' from measurements of some other quantity. In Figure 2.7, some of the outputs can be measured and used directly in the feedback control; other outputs required by the controller cannot be measured directly so some other process measurement is made and an estimator used to calculate what the controlled variable will be. Inferential measurements are frequently used in distillation column control. A schematic of a binary distillation column is shown in Figure 2.8. The five independent variables which are to be controlled are usually the liquid levels H_a and H_b in the accumulator and the reboiler, and the compositions X_a and X_b of the top and bottom products. The compositions can be measured directly by spectographic techniques but it is more usual to measure the temperatures at points Y_a and Y_b near the top and bottom of the column. The temperatures represent the boiling points of the mixture at the position in the column and from measurements of pressure and temperature the compositions can be inferred [Edwards 1982].

Feedforward control is frequently used in the process industries; it involves measuring the disturbances on the system rather than measuring the outputs and is illustrated in Figure 2.9. For example, in the rolling of sheet steel, if the temperature of the billet is known as it approaches the first-stage mill, the initial setting of the roll gap can be calculated accurately and estimates of the reduction at each stage of the mill can be made; hence the initial gaps for the subsequent stages can also be calculated. If this is done the time taken to get the gauge of the steel within tolerance can be much reduced and hence the quantity of scrap (out of

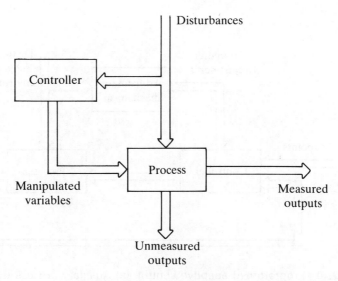

Fig. 2.9 General structure of a feedforward control configuration.

tolerance) steel reduced. The effect of introducing feedforward control is to speed up the response of the system to disturbances; it can, however, only be used for disturbances which can be measured.

Adaptive control can take several forms; three of the most common are:

- programmed adaptive control;
- self-tuning; and
- model-reference adaptive control.

Programmed adaptive control is illustrated in Figure 2.10. The adaptive, or adjustment, mechanism makes preset changes on the basis of changes in either

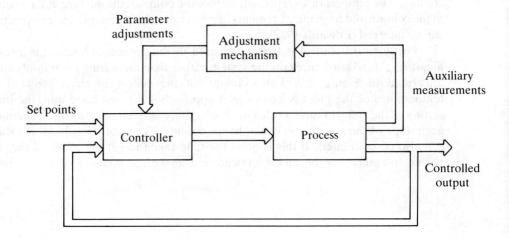

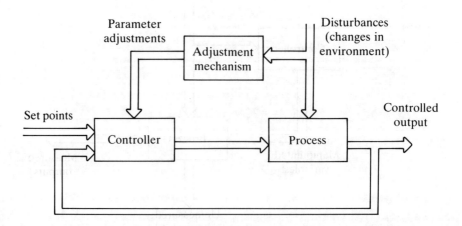

Fig. 2.10 Programmed adaptive control. (a) Auxiliary process measurements; and (b) external environment.

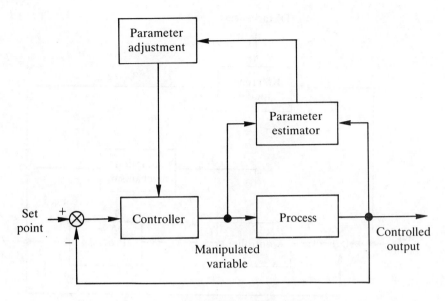

Fig. 2.11 Self-tuning adaptive control.

auxiliary process measurements (Figure 2.10a) or the external environment (Figure 2.10b). The latter method is used in aircraft controls, for example, when control parameters are changed with alterations in altitude and aircraft speed. Adaptive control using self-tuning is illustrated in Figure 2.11 and uses identification techniques to achieve continual determination of the parameters of the process being controlled; changes in the process parameters are then used to adjust the actual controller. The model reference technique is illustrated in Figure 2.12; it relies on the ability to construct an accurate model of the process and to measure the disturbances which affect the process.

2.4 SUPERVISORY CONTROL

The adoption of computers for process control has increased the range of activities that can be performed, for not only can the computer system directly control the operation of the plant, it can also provide managers and engineers with a comprehensive picture of the status of the plant operations. It is in this *supervisory* role and in the presentation of information to the plant operator – large rooms full of dials and switches have been replaced by VDUs and keyboards – that the major changes have been made; the techniques used in the basic feedback control of the plant have changed little from the days when pneumatically operated three-term

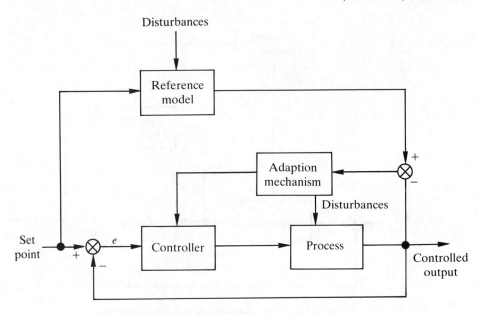

Fig. 2.12 Model-reference adaptive control.

controllers were the norm. Direct digital control (DDC) is often simply the computer implementation of the techniques used for the traditional analog controllers.

Many of the early computer control schemes used the computer in a supervisory role and not for DDC. The main reasons for this were that computers in the early days were not always very reliable and caution dictated that the plant should still be able to run in the event of a computer failure, and that computers were very expensive and it was not economically viable to use a computer to replace the analog control equipment in current use. A computer system used to adjust the set points of the existing analog control system in an optimum manner (to minimize energy or to maximize production) could perhaps be economically justified (see Figure 2.13).

A simple example of a system in which supervisory control might be used is shown in Figure 2.14. Two evaporators are connected in parallel and material in solution is fed to each unit. The purpose of the plant is to evaporate as much water as possible from the solution. Steam is supplied to a heat exchanger linked to the first evaporator and the steam for the second evaporator is supplied from the vapours boiled off from the first stage. To achieve maximum evaporation the pressures in the chambers must be as high as safety permits. However, it is necessary to achieve a balance between the two evaporators: if the first is driven at its maximum rate it may generate so much steam that the safety thresholds for the second evaporator are exceeded. A supervisory control scheme could be designed to balance the operation of the two evaporators to obtain the best overall evaporation rate.

Most applications of supervisory control are very simple and are based upon knowledge of the steady state characteristics of the plant. In a few systems complex

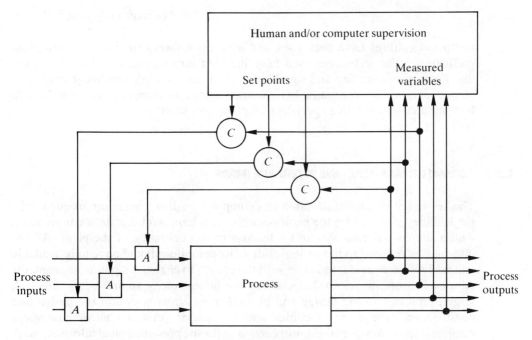

Fig. 2.13 Supervisory control.

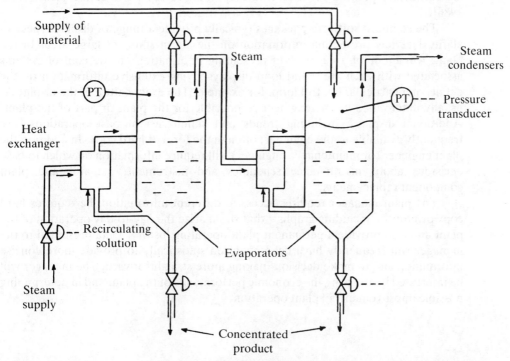

Fig. 2.14 An evaporation plant (redrawn from Bennett and Linkens, *Real-time Computer Control*, 1984, Peter Peregrinus).

control algorithms have been used and have been shown to give increased plant profitability. The techniques used have included optimization based on hill-climbing, linear programming and simulations involving complex non-linear models of plant dynamics and economics. In these applications the complex algorithms have to be computed in real time in parallel with plant operation.

2.5   HUMAN OR MAN-MACHINE INTERFACE (MMI)

The key to the successful adoption of a computer control scheme can frequently be the facilities provided for the plant operator. It is important that he/she is provided with a simple and clear system for the day-to-day operation of the plant. All the information relevant to the current state of its operation should be readily available and facilities to enable interaction with the plant – to change set points, to manually adjust actuators, to acknowledge alarm conditions, etc. – should be provided. A large proportion of the design and programming effort goes into the design and construction of operator facilities and the major process control equipment companies have developed extensive schemes for the presentation of information. A typical operator station has specially designed keyboards and several display and printer units; extensive use is made of color displays and mimic diagrams; video units are frequently provided to enable the operator to see parts of the plant [Jovic 1986].

The standard software packages typically provide a range of display types: an alarm overview presenting information on the alarm status of large areas of the plant; a number of area displays presenting information on the control systems associated with each area; and loop displays giving extensive information on the details of a particular control loop, for example. The exact nature of the displays is usually determined by the engineer responsible for the plant or part of the plant. Additional displays, including trends and summaries of past operations, are frequently available to the engineer, often in the form of hard copy. In addition the plant engineer (or maintenance engineer) will require information on which to base decisions about maintenance schedules and instrument, actuator and plant component replacements.

The plant manager requires access to different information: he requires hard copy printouts – including graphs – that summarize the day-to-day operation of the plant and also provide a permanent plant operating history. Data presented to the manager will frequently have been analyzed statistically to provide more concise information and to make decision-making more straightforward. The manager will be interested in assessing the economic performance of the plant and in determining possible improvements in plant operation.

2.6 THE CONTROL ENGINEER

Assuming that a decision has been made on the most suitable computer system, the control engineer's responsibility is as follows:

1. To define the measurements and actuations and to set up scaling and filter constants, alarm and actuator limits, sampling intervals, etc.
2. To define the DDC controllers, the interlinking or cascading of such controllers and the connections with any other elements in the control scheme.
3. To tune the control scheme, i.e., to select the appropriate gains so that they perform according to some desired specification.
4. To define and program the sequence control procedures necessary for the automation of plant operation.
5. To determine and implement satisfactory supervisory control schemes.

For a large project all the above requirements are too great for any one person to handle and in such cases a team of engineers would be involved. Additionally, if the programming of the system had to be done from scratch for each individual case, the task would be very burdensome and costly. However, process control applications have many features in common and the major suppliers of process control computer systems offer an extensive range of software packages, so that for each application the main task of the engineer/programmer is to select the appropriate modules and to assemble the required database describing the particular plant. The major omission of much of the standard software is that it fails to provide assistance with the tuning of the plant.

Software for the supervisory and sequencing operations has to be much more flexible in that the range of actions that may be required for any plant is much greater than for the DDC control part. Frequently, it is necessary to program this part of the system specifically for each plant.

With the increasing use of microprocessors in a wide variety of applications the standard software approach used by the process controllers is becoming inadequate. It has been based largely on the requirements of large plants with long time-constants, with a single large computer, but the move is now towards systems in which a computer or computers are *embedded* as part of the plant or as a piece of equipment and which are programmed to carry out only those functions required, i.e., they do not have within them general purpose software, a large proportion of which may be irrelevant to the particular application. This development has changed the approach to the design of real-time computer control systems and made it necessary for the control engineer to have a greater knowledge of programming languages and operating systems.

2.7 CENTRALIZED COMPUTER CONTROL

Throughout most of the 1960s computer control implied the use of one central computer for the control of the whole plant. The reason for this was largely financial: computers were expensive. From the previous sections it should now be obvious that a typical computer-operated process involves the computer in performing many different types of operations and tasks. Although a general purpose computer can be programmed to perform all of the required tasks the differing time-scales and security requirements for the various categories of task makes the programming job difficult, particularly with regard to the testing of software. For example, the feedback loops in a process may require servicing at intervals measured in seconds while some of the alarm and switching systems may require a response in less than one second; the supervisory control calculations may have to be repeated at intervals of several minutes or even hours; production management will want summaries at shift or daily intervals; and works management will require weekly or monthly analyses. Interrelating all the different time-scales can cause serious difficulties.

A consequence of centralized control was the considerable resistance to the use of DDC schemes in the form shown in Figure 2.4: with one central computer in the feedback loop, failure of the computer would result in the loss of control of the whole plant. In the 1960s computers were not very reliable: mean time-to-failure was typically 3 to 6 months. Many of the early schemes were therefore for supervisory control as shown in Figure 2.13.

However, in the mid-1960s the traditional process instrument companies began to produce digital controllers with analog back-up. These units were based on the standard analog controllers but allowed a digital control signal from the computer to be passed through the controller to the actuator; the analog system tracked the signal and if the computer did not update the controller within a specified (adjustable) interval the unit dropped on to local analog control. This scheme enabled DDC to be used with the confidence that if the computer failed, the plant could still be operated. The cost, however, was high in that two complete control systems had to be installed.

By 1970 the cost of computer hardware had reduced to such an extent that it became feasible to consider the use of dual computer systems (Figure 2.15). Here, in the event of the failure of one of the computers, the other takes over. In some schemes the changeover is manual, in others automatic failure detection and changeover is incorporated. Many of these schemes are in use. They do, however, have a number of weaknesses: cabling and interface equipment is not usually duplicated, neither is the software in the sense of having independently designed and constructed programs, so that the lack of duplication becomes crucial. Automatic failure and changeover equipment when used becomes in itself a critical component. Furthermore, the problems of designing, programming, testing and maintaining the software are not reduced; if anything they are further complicated in that provision for monitoring ready for changeover has to be provided.

The continued reduction of the cost of hardware and the development of the microprocessor has made multicomputer systems feasible. These fall into two types:

Hierarchical Systems

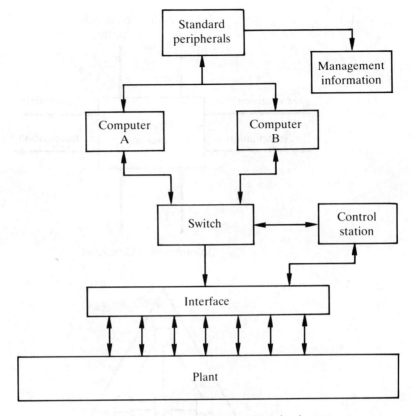

Fig. 2.15 Dual computer control scheme.

1. Hierarchical Tasks are divided according to function, e.g. with one computer performing DDC calculations and being subservient to another which performs supervisory control.
2. Distributed Many computers perform essentially similar tasks in parallel.

2.8 HIERARCHICAL SYSTEMS

This is the most natural development in that it follows both the typical company decision-making structure shown in the pyramid in Figure 2.16: each decision element receives commands from the level above and sends information back to that level and, on the basis of information received from the element or elements below and from constraints imposed by elements at the same level, sends commands to the element(s) below and information to element(s) at the same level. This structure also follows a natural division of the production process in terms of the time-response

(a)

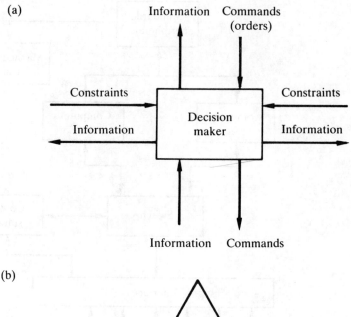

(b)

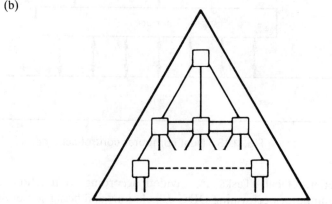

Fig. 2.16 Hierarchical decision-making. (a) Decision-making function; (b) decision-making structure.

requirements of the different levels. At the bottom of the pyramid, or hierarchy, fast response (measured in milliseconds or seconds) to simple problems is required: as one progresses up the hierarchy the complexity of the calculations increases as does the time allowed for the response.

A typical example of a hierarchical system is the batch system shown in Figures 2.17a and 2.17b. This system has three levels which we have called manager, supervisor and unit control. It is assumed that single computers are used for the manager and supervisor functions and that for each processing unit a single unit control computer is used. At the manager level functions such as resource allocation, production scheduling and production accounting are carried out. The input information may be, for example, sales orders (actual and forecast), stock

Hierarchical Systems

levels, selling cost and production costs (or profit margins) on each product, operating costs for each process unit and scheduled maintenance plans for each operating unit, and the current state of production units. On the basis of this information the production schedule, i.e., the list of products to be produced and the quantities and process unit to be used, will be calculated. This may be done daily or at some other interval depending on production times, etc.

The information regarding the production schedule is transferred to the supervisor. It is assumed that the supervisor has a store containing the product recipe (i.e., how to make a particular product) and a store of the operation sequences for making the product. When the appropriate unit, as selected by the production plan, is ready the information on the product – set points, alarm conditions, tolerances, etc. – is down loaded into the unit controller as are the particular operations to be

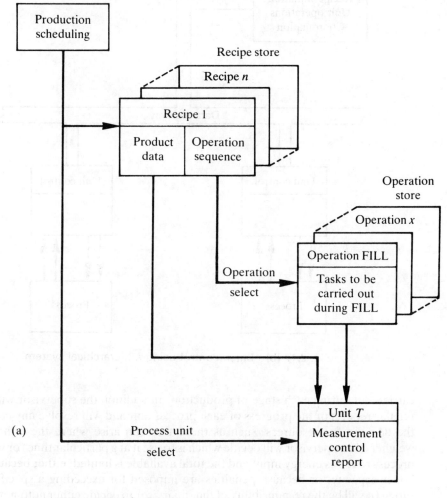

Fig. 2.17 (a) Batch control using a hierarchical system.

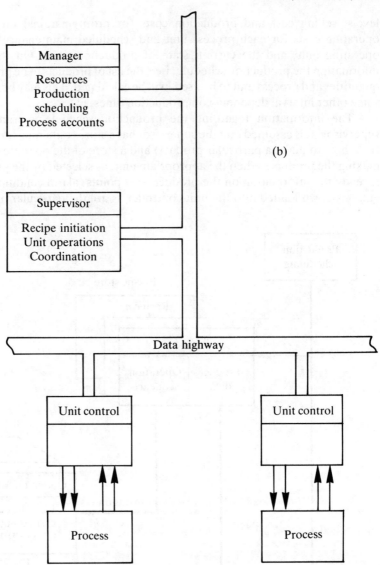

Fig. 2.17 (b)   Batch control using a hierarchical system.

carried out during each stage of production. In addition the supervisor will receive regular reports of the progress of each process unit and will resolve any conflicts in the demand for resources, such as those that can arise where the units share a weigher (the supervisor will decide which is to use it at a particular time) or where the process requires energy input and the total available is limited, either because of the boiler capacity, or because penalties are imposed for exceeding a specified kVA rating (it will be the responsibility of the supervisor to decide either automatically, or

Distributed Systems 43

through a request to a human supervisor, which unit is to reduce consumption or in what proportion consumption is to be reduced).

At the lowest level, the unit controllers are responsible for operating the plant: opening and closing valves and switches; controlling temperatures, pressures, speeds, flows; monitoring alarms; and reporting plant conditions.

Most hierarchical systems will involve some form of distributed network and hence most systems will be a mixture of hierarchical and distributed control.

2.9 DISTRIBUTED SYSTEMS

The underlying assumptions of the distributed approach are that

1. each unit is carrying out essentially similar tasks to all the other units;
2. in the event of failure or overloading of a particular unit all or some of the work can be transferred to other units.

In other words, the work is not divided by function and allocated to a particular computer as in a hierarchical system; instead, the total work is divided up and spread across several computers. This is a conceptually simple and attractive approach – many hands make light work – but it poses difficult hardware and software problems since, in order to gain the advantages of the approach, allocation of the tasks between computers has to be dynamic, i.e., there has to be some mechanism which can assess the work to be done and the present load on each computer in order to allocate work. Because each computer needs access to all the information in the system, high bandwidth data highways are necessary. There has been considerable progress in developing such highways and the various types are discussed below; computer scientists and engineers are also carrying out considerable research on multiprocessor computer systems and this work could lead to totally distributed systems becoming feasible.

There is also a more practical approach to distributing the computing load whereby no attempt is made to provide for dynamic allocation of resources but instead a simple ad hoc division is adopted with, for example, one computer performing all non-plant input and output, one computer performing all DDC calculations, another performing data acquisition and yet another the control of the actuators.

In most modern schemes a mixture of distributed and hierarchical approaches are used as shown in Figure 2.18. The tasks of measurement, DDC, operator communications, etc. are distributed among a number of computers which are linked together via a common serial communications highway and are configured in a hierarchical command structure. For illustration the figure shows five broad divisions of function:

Level 1   All computations and plant interfacing associated with measurement and actuation. This level provides measurement and actuation database for the whole system.

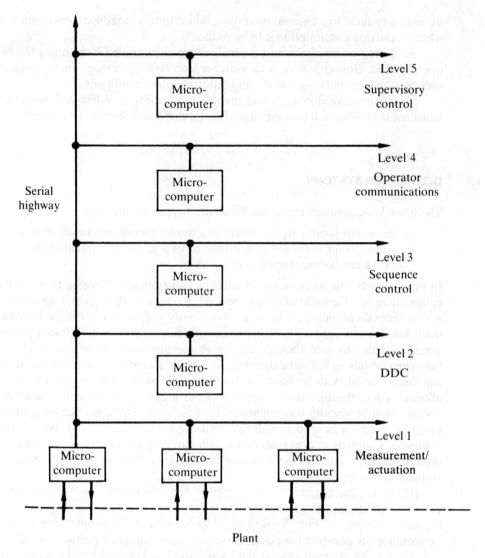

Fig. 2.18 A distributed and hierarchical system.

    Level 2   All DDC calculations.
    Level 3   All sequence calculations.
    Level 4   Operator communications.
    Level 5   Supervisory control.
    Level 6   Communications with other computer systems.

It is not necessary to preserve rigid boundaries, e.g., a DDC unit may perform some sequencing or may interface directly to plant.

The major advantages of this approach are:

1. The system capabilities are greatly enhanced by the sharing of tasks between processors – the burden of computation for a single processor becomes very great if all of the described control features are included. One of the main computing loads is that of measurement scanning, filtering and scaling, not because any one calculation is onerous but because of the large number of signals involved and the frequency at which the calculations have to be repeated. Separation of this aspect from the DDC, even if only into two processors, greatly enhances the number of control loops that can be handled. The DDC computer will collect measurements, already processed, via the communications link at a much lower frequency than that at which the measurement computer operates.
2. The system is much more flexible than the use of a single processor; if more loops are required or an extra operator station is needed, all that is necessary is to add more boxes to the communication link – of course the other units on the link will need to be updated to be aware of the additional items. It also allows standardization, since it is much easier to develop standard units for well-defined single tasks than for overall control schemes.
3. Failure of a unit will cause much less disruption in that only a small portion of the overall system will not be working. Provision of automatic or semi-automatic transfer to a back-up system is much easier.
4. It is much easier to make changes to the system, either in the form of hardware replacements or software changes. Changing large programs is hazardous because of the possibility of unforeseen side-effects; with the use of small modules such effects are less likely to occur and are more easily detected and corrected.
5. Linking by serial highway means that the computer units can be widely dispersed; hence it is unnecessary to bring cables carrying transducer signals to a central control room.

2.10 ECONOMICS OF COMPUTER CONTROL SYSTEMS

Before the widespread availability of microprocessors, computer control was expensive and a very strong case was needed to justify the use of computer control rather than conventional instrumentation. In some cases computers were used because otherwise plant could not have been made to work profitably; this is particularly the case with large industrial processes that require complex sequencing operations. The use of a computer permits the repeatability that is essential in plants used for the manufacture of drugs, say. In many applications flexibility is important – it is difficult with conventional systems to modify the sequencing procedure to provide for the manufacture of a different product. Flexibility is particularly important when the product or the product specification may have to be changed frequently; with a computer system it is simple to maintain a database containing the product recipes and thus to change to a new recipe quickly and reliably.

The application of computer control systems to many large plants has frequently been justified on the grounds that even a small increase in productivity (say 1 or 2%) will more than pay for the computer system. After installation it has frequently been difficult to established that an improvement has been achieved, sometimes production has decreased, but the computer proponents have then argued that but for the introduction of the computer system production would have decreased by a greater amount!

Some of the major benefits to accrue from the introduction of a computer system have been in the increased understanding of the behavior of the process that has resulted from the studies necessary to design the computer system and from the information gathered during running. This has enabled supervisory systems to be designed to keep the plant running at an operating point closer to the desired point. The other main area of benefit has been in the control of starting and stopping of batch operations in that computer based systems have generally significantly reduced the dead time associated with batch operations.

The economics of computer control has been changed drastically by the microprocessor in that the reduction in cost and the improvement in reliability has meant that computer-based systems are the first choice in many applications. Indeed, microprocessor based instrumentation is frequently cheaper than the equivalent analog unit. The major costs of computer control are now no longer the computer hardware, but the system design costs and the cost of software; as a consequence attention is shifting towards greater standardization of design of software products and to the development of improved techniques for design (particularly software design) and for software construction and testing.

EXERCISES

2.1 You are the manager of a plant which can produce ten different chemical products in batches which can be between 500 and 5000 kg. What factors would you expect to consider in calculating the optimum batch size? What arguments would you put forward to justify the use of an online computer to calculate optimum batch size?

2.2 What are the advantages/disadvantages of using a continuous oven? How will the control of the process change from using a standard oven on a batch basis and using an oven in which the batch passes through on a conveyor belt? Which will be the easier to control?

REFERENCES AND BIBLIOGRAPHY

BENNETT, S. and LINKENS, D.A. (eds.) (1982), *Computer Control of Industrial Processes*, Peter Peregrinus, Stevenage

BENNETT, S. and LINKENS, D.A. (eds.) (1984), *Real-time Computer Control*, Peter Peregrinus, Stevenage

CASSELL, D.A. (1983) *Microcomputers and Modern Control Engineering*, Prentice Hall
DESHPANDE, P.B. and ASH, R.H. (1983), *Elements of Computer Process Control*, Prentice Hall
EDWARDS, J.B. (1982), 'Process control by computer', in Bennett, S. and Linkens, D.A. (eds.) (1982)
HENRY, R.M. (1987), 'Generating sequences', in Linkens, D.A. and Virk, G.S. (eds.) (1987)
HOLLAND, R.C. (1983), *Microcomputers for Process Control*, Pergamon
JOVIC, F. (1986), *Process Control Systems: Principles of Design and Operation*, Kogan Page
KISSELL, T.E. (1986), *Understanding and Using Programmable Controllers*, Prentice Hall
LEIGH, J.R. (1985), *Applied Digital Control*, Prentice Hall
LINKENS, D.A. and VIRK, G.S. (eds.) (1987), *Computer Control*, Institute of Measurement and Control for SERC
LOWE, E.I. and HIDDEN, A.E. (1971) *Computer Control in Process Industries*, Peter Peregrinus, Stevenage
LUYBEN, W.L. (1974), *Process Modelling, Simulation and Control for Chemical Engineers*, McGraw-Hill
MELLICHAMP, D.A. (ed.) (1983), *Real-time Computing with Applications to Data Acquisition and Control*, Van Nostrand
SANDOZ, D.J. (1984), 'A survey of computer control' in Bennett, S and Linkens, D.A. (eds.) (1984)
SAVAS, E.S. (1965), *Computer Control of Industrial Processes*, McGraw-Hill
STEPHANOPOULOS, G. (1985), *Chemical Process Control*, Prentice Hall

3

Computer Hardware Requirements for Real-time Applications

3.1 INTRODUCTION

Although almost any digital computer can be used for real-time computer control and other real-time operations, they are not all equally easily adapted for such work. A process control computer has to communicate both with plant and personnel: this communication must be efficient and effective and the processor must be capable of rapid execution to provide for real-time control action.

A characteristic of computers used in control systems is that they are modular: they provide the means of adding extra units, in particular, specialized input and output devices, to a basic unit. The capabilities of the basic unit, in terms of its processing power, storage capacity, input/output bandwidth and interrupt structure, determine the overall performance of the system. A simplified block diagram of the basic unit is shown in Figure 3.1; the arithmetic and logic, control, register, memory and input/output units represents a general purpose digital computer.

Of equal importance in a control computer are the input/output channels which provide a means of connecting process instrumentation to the computer, and also the displays and input devices provided for the operator. The instruments are not usually connected directly but by means of interface units; a typical range of such units is shown in Figure 3.2. Also of importance is the ability to communicate with other computers since, as we discussed in the previous chapter, many modern computer control systems involve the use of several interconnected computers.

3.2 GENERAL PURPOSE COMPUTER

3.2.1 Central processing unit

The Arithmetic and Logic Unit (ALU) together with the control unit and the general purpose registers make up the central processing unit (CPU). The ALU contains the circuits necessary to carry out arithmetic and logic operations, for example, to add

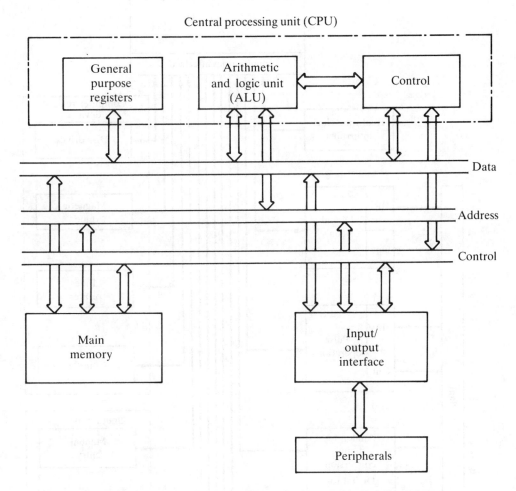

Fig. 3.1 Schematic diagram of a general purpose digital computer.

numbers, subtract numbers and compare two numbers. Associated with it may be hardware units to provide multiplication and division of fixed point numbers and, in the more powerful computers, a floating point arithmetic unit. The general purpose registers can be used for storing data temporarily while it is being processed. Early computers had a very limited number of general purpose registers and hence frequent access to main memory was required. Most computers now have CPUs with several general purpose registers – some large systems have as many as 256 registers – and for many computations intermediate results can be held in the CPU without the need to access main memory thus giving faster processing.

The control unit continually supervises the operations within the CPU: it fetches program instructions from main memory, decodes the instructions and sets up the necessary data paths and timing cycles for the execution of the instructions.

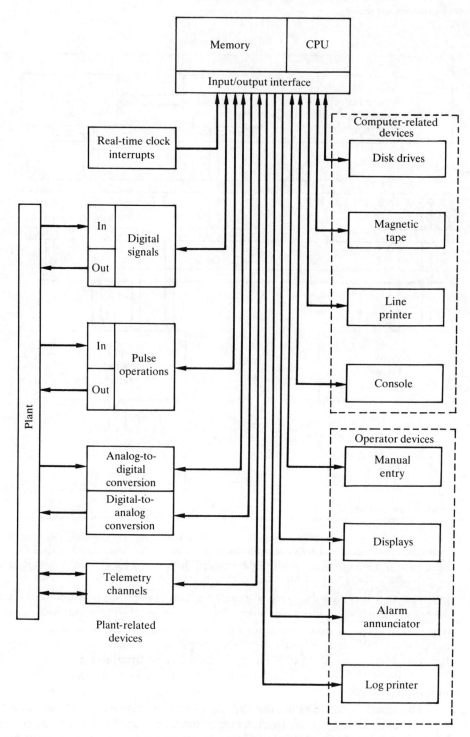

Fig. 3.2 Typical interface devices.

General Purpose Computer

The important features of the CPU which determine the processing power available and hence influence the choice of computer for process control include:

- word length;
- instruction set;
- addressing methods;
- number of registers;
- information transfer rates; and
- interrupt structure.

The word length used by the computer is important both in ensuring adequate precision in calculations and in allowing direct access to a large area of main storage within one instruction word. It is always possible to compensate for short word lengths, both for arithmetic precision and for memory access, by using multiple word operations, but the penalty is increased time for the operations. The amount of main storage directly accessible is also influenced by the number of address lines provided on the input/output interface. Some indication of the inherent precision available from a given word length (and also the directly addressable memory) is given in Table 3.1.

Table 3.1

Word length	Integer range	Memory size
8	-128 to $+127$	256
16	$-32\,768$ to $+32\,767$	64K
32	$-4\,294\,967\,296$ to $+4\,294\,967\,295$	8M

Formula is integer range $= -2^n$ to $+2^n - 1$ where $n =$ number of bits in a word.

The basic instruction set of the CPU is also important in determining its overall performance. Features which are desirable are:

- flexible addressing modes for direct and immediate addressing;
- relative addressing modes;
- address modification by use of index registers;
- instructions to transfer variable length blocks of data between storage units or locations within memory; and
- single commands to carry out multiple operations.

These features reduce the number of instructions required to perform 'housekeeping' operations and hence both reduce storage requirements and improve overall speed of operation by reducing the number of accesses to main memory required to carry out the operations. A consequence of an extensive and powerful instruction set is, however, that efficient programming in the assembly language becomes more difficult in that the language can become complex: thus it is desirable to be able to program the system using a high-level language which has a compiler designed to make optimum use of the special features of the instruction set.

Another area which must be considered carefully when selecting a computer for process control is information transfer, both within the CPU, between backing store and the CPU, and with the input/output devices. The rate at which such transfers can take place, the ability to carry out operations in parallel with processing of data and the ability to communicate with a large range of devices can be crucial to the application to process control. A vital requirement is also a flexible and efficient multilevel interrupt structure.

3.2.2 Storage

The storage used on computer control systems divides into two main categories: fast access storage and auxiliary storage. The fast access memory is that part of the system which contains data, programs, and results which are currently being operated on. On early computer systems, because of the high cost of fast access memory, such storage was limited to 32 K to 64 K words; now, with the fall in the cost of semiconductor memory, much larger amounts of fast access storage can be provided. The major restriction with current computers is commonly the addressing limit of the processor. In addition to RAM memory it is now common to have ROM, PROM or EPROM memory for the storage of critical code or predefined functions.

The use of ROM has eased the problem of memory protection to prevent loss of programs through power failure or corruption by malfunctioning of the software (this can be a particular problem during testing). An alternative to using ROM is the use of memory mapping techniques that trap instructions which attempt to store in a protected area. This technique is usually only used on the larger systems which use a memory management system to map program addresses onto the physical address space. An extension of the system allows particular parts of the physical memory to be set as read only, or even locked out altogether; write access can be gained only by the use of 'privileged' instructions.

Memory management units have frequently been added to computer systems which use a CPU designed at a time when fast memory was expensive. These CPUs have a limited number of address lines and hence a limited address space. An example of such a range of computers is the PPD-11 series of machines. The PDP-11s were originally limited to a maximum memory of 28 K words but they can now be extended to take a much larger amount of memory through the addition of a memory management unit. The unit works by mapping different areas of the physical memory on to the actual address space of the processor. The directly addressable memory is still limited to 28 K.

The auxiliary storage medium is typically disk or magnetic tape. These devices provide bulk storage for programs or data which are required infrequently at a much lower cost than fast access memory. The penalty is a much longer access time and the need for interface boards and software to connect them to the CPU. Auxiliary or backing store devices operate asynchronously to the CPU and care has to be taken in deciding on the appropriate transfer technique for data between the CPU, fast access memory and the backing store. On a real-time system it is not desirable to make use of the CPU to carry out the transfer as not only is this a slow method but no other computation can take place during transfer. For efficiency of transfer it is sensible to transfer large blocks of data rather than a single word or byte and this can result in the CPU not being available for up to several seconds in some cases.

The approach frequently used is direct memory access (DMA). For this the interface controller for the backing memory must be able to take control of the address and data buses of the computer.

3.2.3 Input and output

The input/output (IO) interface is one of the most complex areas of a computer system; part of the complication arises because of the wide variety of devices which have to be connected and the wide variation in the rates of data transfer. A printer may operate at 300 baud whereas a disc may require a rate of 500 Kbaud. The devices may require parallel or serial data transfers, may require analog-to-digital or digital-to-analog conversion, or conversion to pulse rates.

The IO system of most control computers can be divided into three sections:

- process IO;
- operator IO; and
- computer IO.

It is modern practice that all these devices (a typical range is shown in Figure 3.2) share the same bus system and hence the CPU treats all devices in the same way and all devices have to conform to the bus standard.

3.2.4 Bus structure

A bus is, in physical terms, a collection of conductors which carry electrical signals; it may be tracks on a printed circuit board or the wires in a ribbon cable. Components can be plugged or soldered to the bus conductors. The physical form of the bus represents the *mechanical characteristic* of the bus system. In addition in order to design an interface unit to connect to a given bus it is necessary to know the *electrical characteristics* of the bus: signal levels, loading (e.g. a given line may support one standard TTL load), type of output gates (open-collector, tri-state, etc.).

It is also necessary to know what the electrical signals flowing along the bus conductors represents – the *functional characteristics*. The bus lines can be divided into three functional groups:

- address lines;
- data lines; and
- control and status lines.

These can be thought of as *where, what* and *when*. The address lines provide information on where the information is to be sent (or where it is to be obtained from); the data lines show what the information is; and the control and status lines indicate when it is to be sent.

3.3 PROCESS-RELATED INTERFACES

Instruments and actuators connected to the process or plant can take a wide variety of forms: they may be used for measuring temperatures and hence could use thermocouples, resistance thermometers, thermistors etc.; they could be measuring flow rates and use impulse turbines; they could be used to open valves or to control thyristor-operated heaters. In all these operations there is a requirement to convert a digital quantity, in the form of a bit pattern in a computer word, to a physical quantity, or to convert a physical quantity to a bit pattern. It is not a sensible or economic approach to attempt to design a different interface for each specific type of instrument or actuator and hence we look for some commonality between them. It is found that most devices can be allocated to one of the four categories listed below:

1. *Digital quantities*   These can be either binary, i.e., a valve is open or closed, a switch is on or off, a relay should be opened or closed; or a generalized digital quantity, i.e., the output from a digital voltmeter in BCD or other format.
2. *Analog quantities*   Thermocouples, strain gauges, etc., give outputs which are measured in millivolts; these can be amplified using operational amplifiers to give voltages in the range -10 to $+10$ volts; conventional industrial instruments frequently have a current output in the range 4 to 20 mA (current transmission gives much better immunity to noise than transmission of low voltage signals). The characteristic of these signals is that they are continuous variables and have to be both sampled and converted to a digital value.
3. *Pulses and Pulse Rates*   A number of measuring instruments, particularly flow meters, provide output in the form of pulse trains; similarly the increasing use of stepping motors as actuators requires the provision of pulse outputs. Many traditional controllers have also used pulse outputs; e.g., valves controlling flows are frequently operated by switching a dc or ac motor on and off, the length of the on pulse being a measure of the change in valve opening required.
4. *Telemetry*   The increasing use of remote outstations, e.g., electricity substations and gas pressure reduction stations, has increased the use of telemetry. The data may be transmitted by landline, radio or the public telephone network; it is, however, characterized by being sent in serial form, usually encoded in standard ASCII characters. For small quantities of

Process-related Interfaces 55

data the transmission is usually asynchronous. Telemetry channels may also be used on a plant with a hierarchy of computer systems instead of connecting the computers by some form of network. An example of this is the CUTLASS system used by the Central Electricity Generating Board which uses standard RS232 lines to connect a hierarchy of control computers.

The ability to classify the interface requirements into the above categories means that a limited number of interfaces can be provided for a process control computer. The normal arrangement is to provide a variety of interface cards which can be added to the system to make up the appropriate configuration for the process to be controlled, e.g., a process with a large number of temperature measurements may have to have several analog input boards.

3.3.1 Digital signal interfaces

A simple digital input interface is shown in Figure 3.3. It is assumed that the plant outputs are logic signals which appear on lines connected to the digital input register. It is usual to transfer one word at a time to the computer so normally the digital input register will have the same number of input lines as the number of bits in the computer word. The logic levels on the input lines will typically be 0 and +5 volts; if the contacts on the plant which provide the logic signals use different levels then conversion of signal levels will be required.

In order to read the lines connected to the digital input register the computer has to place on the address bus the address of the register while some decoding circuitry is required in the interface (address decoder) to select the digital input register. In addition to the 'select' signal, an 'enable' signal may also be required; this could be provided by the 'read' signal from the computer control bus. In response to both the 'select' and 'enable' signals the digital input register would enable its output gates to put data onto the computer data bus. Note that for proper operation of the data bus the digital input register must only connect its output gates to the data bus when it is selected and enabled; if it connects at any other time it will corrupt data intended for other devices.

The timing of the transfer of information will be governed by the CPU timing. A typical example is shown in Figure 3.4. It is assumed for this system that the transfer requires three cycles of the system clock, labeled T_1, T_2, and T_3. The address lines begin to change at the beginning of the cycle T_1 and they are guaranteed to be valid by the start of cycle T_2; also at the start of cycle T_2 the READ line becomes active. For the correct read operation the digital input register has to provide stable data at the negative going edge (or earlier) of the clock during the T_3 cycle and the data must remain on the data lines until the negative going edge of the following clock cycle.

56   Computer Hardware Requirements

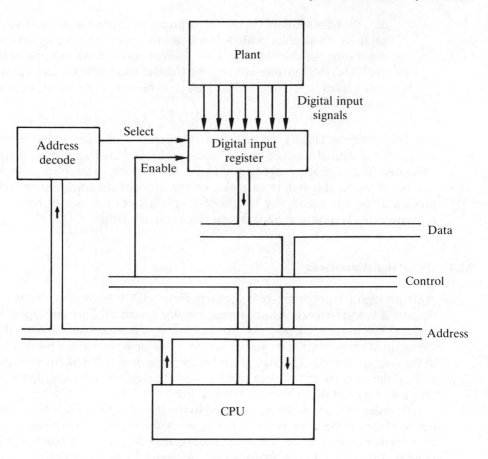

Fig. 3.3 Simple digital input interface.

It should be noted that the actual time taken to transfer the data from the data bus to the CPU may be much shorter than the time for which the data is valid. The requirement that it remain valid from the negative going edge of cycle T_3 until the negative going edge of the following cycle is to provide for the worst case condition arising from variations in the performance of the various components.

The system shown in Figure 3.3 can provide information only on demand from the computer: it cannot indicate to the computer that information is waiting. There are many circumstances in which it is useful to be able to indicate to the computer that the status of one of the input lines has changed. To do this it is necessary to have some form of status line which the computer can test, or which can be used as an interrupt.

A simple digital output interface is shown in Figure 3.5. Digital output is the simplest form of output; all that is required is a register or latch which can hold the data output from the computer. To avoid the data in the register changing when the data on the data bus changes, the output latch must only respond when it is

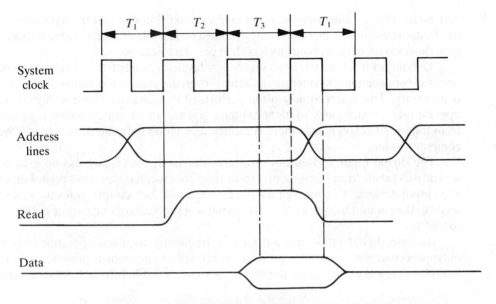

Fig. 3.4 Simplified READ (INPUT) timing diagram.

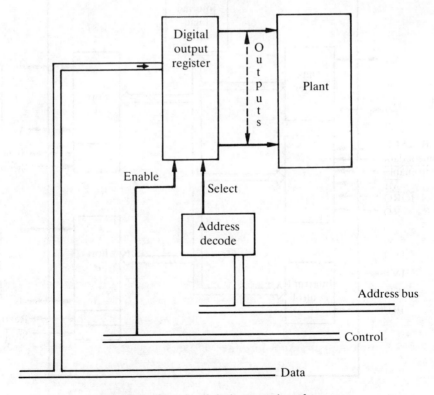

Fig. 3.5 Simple digital output interface.

addressed. The 'enable' signal is used to indicate to the device that the data is stable on the data bus and can be read. The latch must be capable of accepting the data in a very short length of time, typically less than one microsecond.

The output from the latch is a set of logic levels, typically 0 to +5 volts; if these levels are not adequate to operate the actuators on the plant, some signal conversion is necessary. This conversion is often performed by using the low-level signals to operate relays which carry to higher voltage signals; an advantage which is gained from the use of relays is that there is electrical isolation between the plant and the computer system.

The digital input and output interfaces described above can also be used to accept BCD data from instruments, since they are essentially parallel digital input and output devices. A 16-bit digital input device could, for example, transmit 4 BCD digits to the computer (this would correspond to a precision of one part in 10 000 – 0 to 9999).

Because digital input and output is a frequently required operation, many microprocessor manufacturers produce integrated circuits which provide such an interface. Typical examples are the Motorola VIA (Versatile Interface Adapter) and

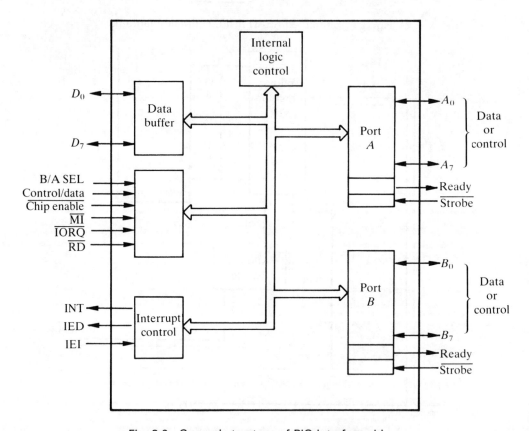

Fig. 3.6 General structure of PIO interface chip.

the Zilog PIO (Parallel Input and Output) devices. The general structure of the PIO is shown in Figure 3.6. The device provides two 8-bit interfaces – *ports* – which can be programmed to operate in several different ways. The ports are referred to as A and B. Port A can operate in four modes:

- Mode 0   The port provides 8 output lines.
- Mode 1   The port provides 8 input lines.
- Mode 2   The port is bidirectional, the direction of individual lines is set by the contents of the data direction register.
- Mode 3   The port operates in control mode.

Port B can operate in Modes 0, 1, and 3 but not in Mode 2; if port A is in Mode 2 then port B must be used in Mode 3.

The various modes are selected by sending control words to the device and hence the interface can be programmed to operate in a variety of ways. In addition to passing data to and from the computer, the PIO can be programmed to provide a variety of status and interrupt signals.

3.3.2 Pulse interfaces

In its simplest form a pulse input interface consists of a counter connected to a line from the plant. The counter is reset under program control and after a fixed length of time the contents are read by the computer. A typical arrangement is shown in Figure 3.7, which also shows a simple pulse output interface. The transfer of data from the counter to the computer uses techniques similiar to those for the digital input described above.

The measurement of the length of time for which the count proceeds can be carried out either by a logic circuit in the counter interface or by the computer. If the timing is done by the computer then the 'enable' signal must inhibit further counting of pulses. If the computing system is not heavily loaded, the external interface hardware required can be reduced by connecting the pulse input to an interrupt and counting the pulses under program control.

Pulse generators can be of two types: they can either send a series of pulses of fixed duration or a single pulse of variable length. For the former, the computer can be used either to turn a pulse generator on or off, or to load a register with the number of pulses to be transmitted. The pulse output is both sent to the process and used to decrement the register contents; when the register reaches zero the pulse output is turned off. A system of this type could be used, e.g., to control the movement of a stepping motor. In the other system the computer can be used to raise or lower a logic line and thus send a variable length pulse to the plant or it can be used to load a register with a number specifying the length of pulse required and the interface logic used to generate the pulse. The variable length pulse system is typically used to operate process control valves.

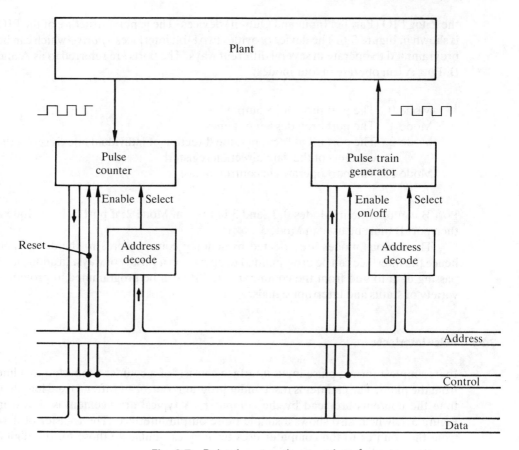

Fig. 3.7 Pulse input and output interface.

Using the computer to turn on or off the pulse train directly is usual only on small systems with a few input or output lines, or when the pulse-rate is low. For large systems or for high pulse rates it is normal to arrange for the interface logic to generate the actual pulse train or to control the duration of the pulse.

Closely related to pulse counters are *hardware timers*. If the pulse counter is made to count down from a preset value at a fixed rate it can act as a timer. For example, if the clock rate used to decrement the counter is set at one count per millisecond, it can be used as a timer with a precision of 1 ms. The method of operation is to load the counter with a binary number corresponding to the desired time interval and to start the count; when the counter reaches zero it generates an interrupt to say that it has 'timed out', i.e. that the interval has elapsed. The computer can then take the appropriate action.

The input to a hardware timer is normally a continuously running accurate pulse generator which may either have a fixed frequency or may be programmable to give a range of frequencies, – thousandths, hundredths, tenths and seconds, for example. The unit is programmed either by setting external switches or by commands sent by the computer. Hardware timers can be used to set the maximum time allowed for the response from an external device: the computer requests a response from a device and at the same time starts a hardware timer; if the device has not responded by the time the hardware timer interrupts then an error condition is generated. A special form of this is the *watchdog timer* which is often used on process control computers. The timer is reset at fixed intervals, usually when the operating system kernel is entered; if the watchdog timer 'times out' it indicates that for some reason the operating system kernel has not been entered at the correct time, either because of some hardware malfunction, or because the normal interrupts have been locked out by a software error. A hardware timer can be used as a real-time clock (see below).

3.3.3 Analog interfaces

The conversion of analog measurements to digital measurements involves two operations: sampling and quantization. The requirements for the sampling rate necessary for controlling a process are discussed in the next chapter. As is shown in Figure 3.8 many analog-to-digital converters (ADC) include a 'sample-hold' circuit on the input to the device. The sample time of this unit is much shorter than the sample time required for the process; this sample hold unit is used to prevent a change in the quantity being measured while it is being converted to a discrete quantity.

The normal method of operation of an analog input interface is that the computer issues a 'start' or 'sample' signal, typically a short pulse (1 microsecond), in response to which the ADC switches the 'sample-hold' into SAMPLE for a short period after which the quantization process commences: quantization may take from a few microseconds to several milliseconds. On completion of the conversion the ADC raises a 'ready' or 'complete' line which is either polled by the computer or is used to generate an interrupt.

Separate ADCs are not normally used for each analog input since, despite the reduction in price in recent years, it would be an expensive approach. Instead, a multiplexer is used to switch the inputs from several input lines to a single ADC, as is shown in Figure 3.8. For high level (0–10V) signals the multiplexer is usually a solid state device (typically based on the use of field effect transistor switches); for low level signals in the millivolt range, e.g., from thermocouples or strain gauges, mercury-wetted reed relay switch units are used. For low level signals, a programmable gain amplifier is usually used between the multiplexer and the sample/hold unit. With a multiplexed system the sequence of operations is more complex than with a single channel device in that the program in the computer has to arrange for the selection of the appropriate input channel.

For the simplest systems a single channel select signal is used which causes the multiplexer to step to the next channel: the channels are thus sampled in sequence. A more elaborate arrangement is to provide channel select input in the form of several

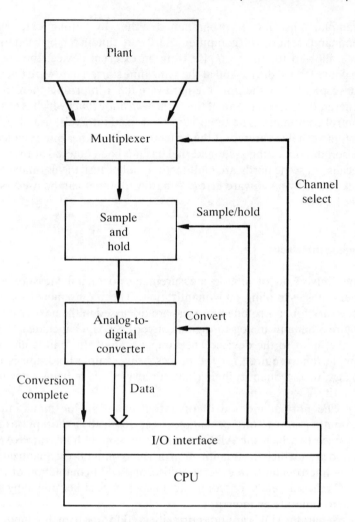

Fig. 3.8 Analog input system.

address inputs which can be connected to the computer data bus. The sequence of events is then: select the channel, send the start conversion command and wait for the conversion complete signal. In some high-speed converters it is possible to send the next channel address during the period in which the present input is being quantized. This technique is also frequently used with reed relay switching since a delay is required between selecting a channel and sampling in order to allow time for the signal to stabilize.

The action of digital-to-analog conversion (DAC) is simpler (and hence cheaper) than ADC and as a consequence it is normal to provide one converter for each output. (It is possible to produce a multiplexer in order to use a single DAC for analog output. Why would this solution not be particularly useful?) A typical arrangement is shown in

Process-related Interfaces 63

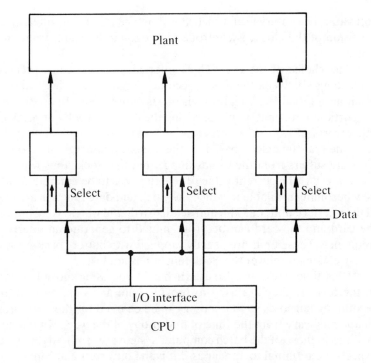

Fig. 3.9 Analog output system.

Figure 3.9. Each DAC is connected to the data bus and the appropriate channel is selected by putting the channel address on the computer address bus. The DAC acts as a latch and holds the previous value sent to it until the next value is sent. The conversion time is typically from 5 to 20 milliseconds and the analog output typically in the range −5 to +5 volts, −10 to +10 volts or a current output 0 to 20 mA.

3.3.4 Real-time clock

A real-time clock is a vital auxiliary device for control computer systems. The hardware unit given the name 'real-time clock' may or may not be a clock; in many systems it is nothing more than a pulse generator with a precisely controlled frequency.

A very common form of clock is based on the use of the ac supply line which is used to generate pulses at 50 (or 60) times per second. By using slightly more complicated circuitry higher pulse rates can be generated, e.g., 100 or 120 pulses per second. The pulses are used to generate interrupts and the software counts the interrupts and hence keeps time. If a greater precision in the time measurement is required a hardware timer is used. A fixed frequency pulse generator (usually crystal-driven) is used to step a down-counter; when it reaches zero it generates an interrupt and reloads the count value. The interrupt activates the real-time clock software. The interval at which the timer generates an interrupt, and hence the precision of the clock, is controlled by loading the count value into the hardware timer.

The choice of the basic clock interval, i.e., the clock precision, has to be a compromise between the timing accuracy required and the load on the CPU. If too small an interval, i.e., high precision, is chosen then the CPU will spend a large proportion of its time simply servicing the clock and will not be able to perform any other work.

The real-time clock based on the use of an interval timer and interrupt-driven software suffers from the disadvantage that the clock stops when the power is lost and on restart the current value of real time has to be entered. Real-time clocks are now becoming available in which the clock and date function are carried out as part of the interface unit, i.e., the unit acts like a digital watch. Real time can be read from the card and the card can be programmed to generate an interrupt at a specified frequency. These units are usually supplied with battery back-up so that even in the absence of mains power the clock function is not lost.

Real-time clocks are also used in batch processing and online computer systems. In the former, they are used to provide date and time on printouts and also for accounting purposes so that a user can be charged for the computer time used: the charge may vary from the time of day or day of the week. In online systems similar facilities to those of the batch computer system are required, but in addition the user expects the terminal to appear as if it is the only terminal connected to the system. The user may expect delays when the program is performing a large amount of calculation but not when it is communicating with the terminal. To avoid any one program causing delays to other programs no program is allowed to run for more than a fraction of a second; typically timings are 200 ms or less. If further processing for a particular program is required it is only performed after all other programs have been given the opportunity to run. This technique is known as time-slicing.

3.4 DATA TRANSFER TECHNIQUES: POLLING

Although the meaning of the data transmitted by the various process, operator and computer peripherals differs, there are many common features which relate to the transfer of the data from the interface to the computer. A characteristic of most interface devices is that they operate synchronously with respect to the computer and that they operate at much lower speeds. This difference in speed would severely limit the speed of operation of the computer if it directly controlled the device; however, for maximum flexibility of operation program control is desirable. Operation in this way is known as 'programmed transfer' and involves the use of the CPU. The alternative is direct memory access (DMA).

With the reduction in cost of integrated circuits and microprocessors, detailed control of the input/output operations is being transferred to IO processors which provide 'buffered entry'. Buffers have been used to collect, e.g., a line of information, before invoking the program requesting the input for a long time in online computing; it is now being extended to the provision of IO processors for real-time systems. For example, an IO processor could be used to control the scanning of a number of analog input channels, only requesting main computer time when it has collected data from all the channels. This could be extended so that the IO processor checked the data to test if any values were outside preset limits set by the main system.

A major problem in data transfer is timing. It may be thought that under programmed transfer, the computer can read or write at any time to a device, i.e., can make an unconditional transfer. For some process output devices, e.g. switches and indicator lights – these would be connected to a digital output interface – or for DACs, unconditional transfer is possible: they are always ready to receive data. For other output devices, e.g., printers and communications channels, which are not fast enough to keep up with the computer but must accept a sequence of data items without missing any item, unconditional transfer cannot be used. The computer must always be sure that the device is ready to accept the next item of data, hence either a timing loop to synchronize the computer to the external device or 'conditional transfer' has to be used. For input, conditional transfer can be used for digital inputs, but not usually for pulse inputs or analog inputs. Where unconditional transfer is used to read the digital value or an analog signal, or the value of a digital instrument rather than simply the pattern of logic indicators, then Gray code or some other form of cyclic binary code should be used to avoid the possibility of large transient errors.

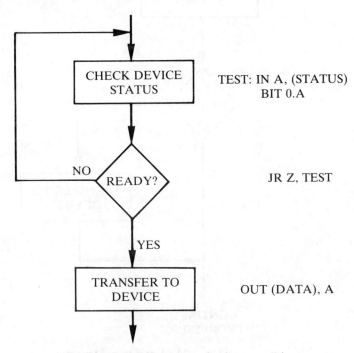

Fig. 3.10 Conditional transfer (busy wait).

A simple example of conditional transfer is shown in Figure 3.10. Assuming that the data is being transferred to a printer which operates at 40 characters per second, the computer will find that the device is ready once every 25 ms. The three instructions involved in performing the test will take approximately 5 microseconds (the actual time will depend on the speed of the processor); thus the conditional test will be carried out about 5000 times for each character transmitted. The computer will spend 99.98% of its time in checking to see if the device is ready and only 0.02% of the time doing useful work: this is clearly inefficient.

As an alternative to providing, on the interface, a status line which can be tested by the computer, a timing loop generated in the computer by loading a register and then decrementing it a specified number of times can be used, e.g.:

```
           LD B,25         ; load register B with time delay
   LOOP:   DEC B           ; decrement B
           JR NZ,LOOP      ; repeat until B is zero
```

To ensure that no transfer is made before the peripheral is ready the time delay must be slightly greater than the maximum delay expected in the peripheral; thus in terms of use of the CPU this method is even more inefficient than the use of the conditional wait. It does slightly simplify and reduce the cost of the interface, however.

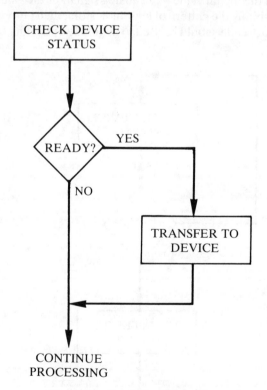

Fig. 3.11 Conditional transfer.

Data Transfer Techniques: Interrupts

An alternative arrangement for conditional transfer, which allows the computer to continue doing useful work if the device is busy, is shown in Figure 3.11.

In this method a check is made to see if the device is ready: if it is ready then the transfer is made; if it is not the computer continues with other work and returns at some later time to check if the device is ready. The technique avoids the inefficiency of waiting in a loop for a device to become ready, but presents the programmer with the difficult task of arranging the software such that all devices are checked at frequent intervals.

The conditional transfer techniques involve *polling*, which is the action of the computer checking whether a device is ready for a data transfer. The problems of polling using conditional waits can be avoided if the computer can respond to an interrupt signal.

3.5 DATA TRANSFER TECHNIQUES: INTERRUPTS

An interrupt is a mechanism by which the flow of the program can be temporarily stopped to allow a special piece of software – an interrupt service routine, or interrupt handler – to run. When this routine has finished, the program which was temporarily suspended is resumed. The process is illustrated in Figure 3.12.

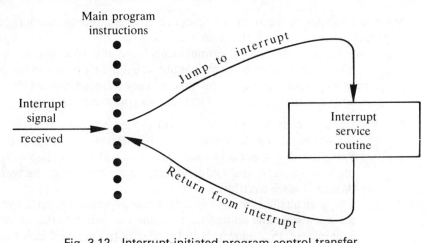

Fig. 3.12 Interrupt-initiated program control transfer.

Interrupts are essential for the correct operation of most real-time computer systems; in addition to providing a solution to the conditional wait problem they are used for:

1. *Real-time clock* The external hardware provides a signal at regularly spaced intervals of time; the interrupt service routine counts the signals and keeps a clock.
2. *Alarm inputs* Various sensors can be used to provide a change in a logic level in the event of an alarm. Since alarms should be infrequent, but may

need rapid response times, the use of an interrupt provides an effective and efficient solution.
3. *Manual override*   Use of an interrupt can allow external control of a system to allow for maintenance and repair.
4. *Hardware failure indication*   Failure of external hardware or of interface units can be signalled to the processor through the use of an interrupt.
5. *Debugging aids*   Interrupts are frequently used to insert breakpoints or traces in the program during program testing.
6. *Operating system*   Interrupts are used to force entry to the operating system before the end of a time slice.
7. *Power failure warning*   It is simple to include in the computer system a circuit that detects very quickly the loss of power in the system and provides a few milliseconds' warning before the loss is such that the system stops working. If this circuit is connected to an interrupt which takes precedence over all other operations in the computer there can be sufficient time to carry out a few instructions which could be sufficient to close the system down in an orderly fashion.

3.5.1 Saving and restoring registers

Since an interrupt can occur at any point in a program, precautions have to be taken to prevent information which is being held temporarily in the CPU registers from being overwritten. All CPUs automatically save the contents of the program counter. This is vital: if the contents are not saved then a return to the point in the program at which the interrupt occurred could not be made. Some CPUs, however, do more and save all the registers. The methods commonly used are:

1. Store the contents of the registers in a specified area of memory. (Note that this implies that an interrupt cannot be interrupted – see below.)
2. Store the registers on the memory stack. This is a simple, widely used method which permits multilevel interrupts; the major disadvantage is the danger of stack overflow.
3. Use of an auxiliary set of registers. Some processors provide two sets of the main registers and an interrupt routine can switch to the alternate set (an example of this approach is the Z80). If only two sets are provided multilevel interrupts cannot be handled. An alternative method is to use a designated area of memory as the working registers and then an interrupt only requires a pointer to be changed to change the working register set; this approach is used in the Texas 9900 series processors.

The use of automatic storage of the working registers is an efficient method if all registers are to be used; it is inefficient if only one or two will be used by the interrupt routine. For this reason fully automatic saving is usually restricted to CPUs with only a few working registers in the CPU; systems with many working registers provide an option either to save or not to save, e.g., the Motorola 6809 which has a normal

Data Transfer Techniques: Interrupts   69

interrupt with automatic saving of registers and a fast interrupt with no automatic saving of registers. Unless response time is critical it is good engineering practice to save all registers; in this way there is no danger that a subsequent modification to the interrupt service routine resulting in a register, which is not being saved, being used. The resulting error would be difficult to find since it would cause random malfunctioning of the system.

The machine status must of course be restored on exit from the interrupt routine; this is straightforward for all methods except for the method which uses the stack to save registers: in this case the registers are restored in the opposite order to that in which they were saved. Systems providing automatic saving also provide automatic restore on exit from the interrupt.

An example of the framework of a Z80 interrupt service routine is shown below.

```
INT1:       CALL SAVREG      ; SAVREG is routine which saves
                             ; working registers
;
; code for interrupt handling is inserted here
;
            CALL RESREG      ; RESREG is routine which restores
                             ; working registers
            EI               ; enable interrupts
            RETI             ; return from interrupt routine
```

The above routine is suitable for a system in which interrupts are not allowed to be interrupted; hence the EI instruction which enables interrupts is not executed until immediately prior to the return from interrupt. The return from an interrrupt routine has to be handled with care to prevent unwanted effects. For example the EI instruction does not re-enable interrupts until after the execution of the instruction which immediately follows it. Therefore, by using the EI/RETI combination a pending interrupt cannot take affect until after the return from the previous one has been completed. It should be noted that the Z80 automatically disables interrupts on acknowledgement of an interrupt and they remain disabled until an EI instruction is executed. Some CPUs operate by disabling interrupts for only one instruction following an interrupt acknowledge; it thus becomes the responsibility of the programmer to disable interrupts as the first instruction of the interrupt service routine if only a single level of interrupt is to be provided.

3.5.2 Interrupt input mechanisms

A simple form of interrupt input is shown in Figure 3.13. In between each instruction the CPU checks the IRQ line. If it is active, an interrupt is present and the 'interrupt service' routine is entered; if it is not active the next instruction is fetched and the cycle repeats. Note that an instruction involves more than one CPU clock cycle and that the interrupt line is checked only between instructions. Because several clock cycles may elapse between successive checks of the interrupt line, the interrupt signal must be latched and only cleared when the interrupt is acknowledged.

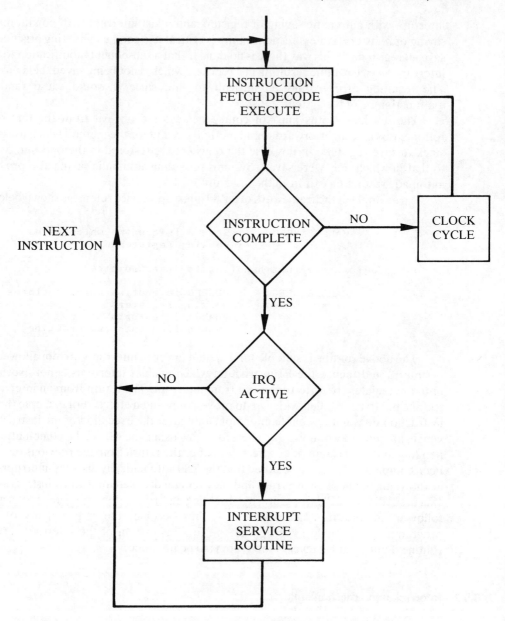

Fig. 3.13 Flowchart of basic interrupt mechanism.

A common arrangement is to have two interrupt lines as shown in Figure 3.14; one of the lines, IRQ, can be enabled and disabled using software and hence the computer can run in a mode in which external events cannot disturb the processing. A second interrupt line is provided; this interrupt cannot be turned off by software

Data Transfer Techniques: Interrupts 71

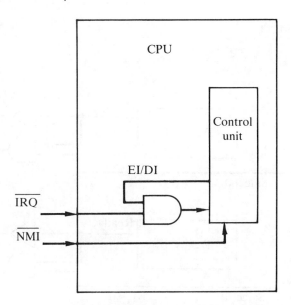

Fig. 3.14   Typical basic interrupt system.

and hence it is said to be a non-maskable interrupt (NMI). A typical use would be to provide the power-failure detect interrupt.

Although most modern computer CPUs have only one or two interrupt lines, a large number of interrupts can be connected by means of an OR gate. It then becomes a problem to determine which of the many external interrupt lines has generated the CPU interrupt.

3.5.3   Interrupt response mechanisms

The CPU may respond to the interrupt in a variety of ways; some of the more popular methods are given below:

1. Transfer control to a specified address – usually in the form of a 'call' instruction.
2. Load the program counter with a new value from a specified register or memory location.
3. Execute a 'call' instruction but to an address supplied from the external system.
4. Use an output signal – an Interrupt Acknowledge – to fetch an instruction from an external device.

Methods 1 and 2 are said to be *software-biased*, in that they require little in the way of external hardware and rely instead on software to determine the interrupt source and the appropriate interrupt service routine. Methods 3 and 4 are *hardware-biased* in that they require more external hardware but can identify the

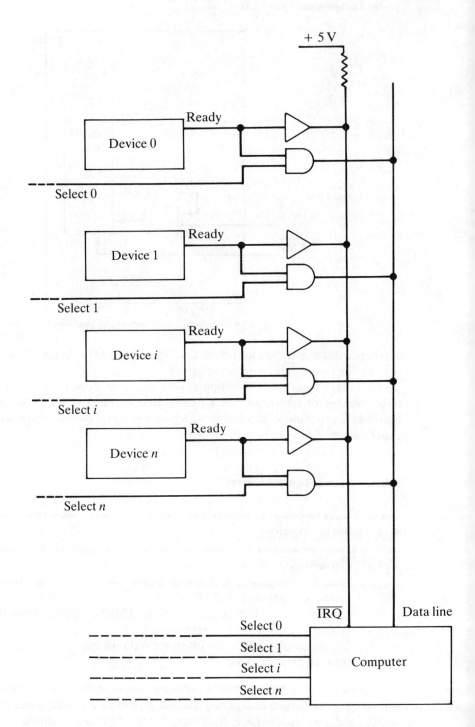

Fig. 3.15 Interrupt system with polling to determine device.

Data Transfer Techniques: Interrupts

interrupt source and can transfer program control directly to the appropriate interrupt service routine.

An example of a system using Method 1 is shown in Figure 3.15. The Z80 microprocessor can be set to operate in three different interrupt modes under program control; if it is set to operate in Mode 1, a fixed interrupt response address, 0038H, is used. If any of the devices become READY, an interrupt will be generated. The CPU will respond by saving the current contents of the program counter and inserting instead the interrupt response address, 0038H. The next instruction to be executed is thus the instruction located at 0038H. This will usually be a jump instruction which will transfer control to the interrupt response routine (IRR). This is shown in Figure 3.16. Control cannot be transferred directly to the interrupt service routine (ISR), since at this stage the actual device generating the interrupt is not known. The interrupt response routine has to poll each device in turn to determine if it is the device which has caused the interrupt. It does this by

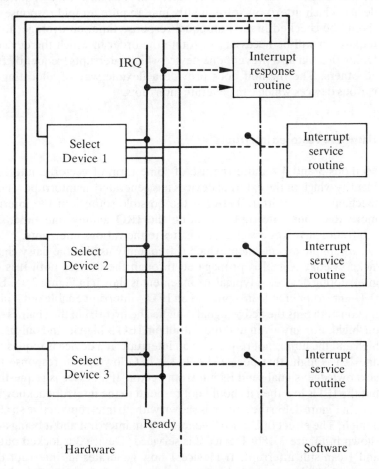

Fig. 3.16 Interrupts, interrupt response routine and interrupt service routines.

addressing each device, reading the data bus and testing bit 0; if bit 0 is set, the device is ready and data transfer can take place. The data transfer would be handled by the interrupt service routine and hence the interrupt response routine would have to transfer control to the appropriate ISR. An outline of the software is given in Figure 3.17.

An example of a device which uses Method 2 is the Motorola 6800 which stores the address of the interrupt response routine in memory locations FFF8H and FFF9H; the address stored here is called the *interrupt vector*. Once the interrupt is detected control is passed to an interrupt response routine and polling must be used to determine which device has caused the interrupt.

The use of polling in interrupt systems has the advantage over normal polling systems that at least one of the inputs is guaranteed to be active. It is clearly, however, not a very satisfactory system if large numbers of devices have to be checked. The load can be reduced by testing the devices which interrupt most frequently first, but this may conflict with response time requirements in that a device which interrupts infrequently may require a rapid response time and hence should be checked first. If an equal response time, on average, for each device is required it will be necessary to rotate the order in which the devices are checked. Doing this can also prevent one device which interrupts frequently from locking out all others. The method does provide a flexible way of allocating priority to the various devices which can generate interrupts.

3.5.4 Hardware vectored interrupts

Methods 3 and 4 require the use of some form of vectored interrupt structure to identify which of the external devices has generated an interrupt. They also require a mechanism to arbitrate between the possible sources of the interrupt to prevent more than one interrupt activating the IRQ at any one time. The process of arbitration involves assigning priorities to the various interrupts.

A simple arrangement which is frequently used is the daisy chain in which the 'acknowledge' signal is propagated through the devices until it is blocked by the interrupting device. A typical arrangement is shown in Figure 3.18. Each unit has an IEI – interrupt enable in – pin and an IEO – interrupt enable out – pin, it is assumed that on both pins the active signal is high. The first IEI in the chain is set permanently on 'high'. For any given unit the output pin IEO is high if, and only if, the input IEI is high and the unit is not requesting an interrupt. If a device is requesting an interrupt and IEI is high, that device should set IEO low and in response to an 'interrupt acknowledge' signal send its interrupt vector. If a device is requesting an interrupt but the IEI is low then it should not respond to the interrupt acknowledge signal.

In Figure 3.18a the system is shown with no interrupts active so all the signals are at high. The effect of Device 2 generating an interrupt and it being acknowledged is shown in Figure 3.18b; Device 2 is serviced, Device 3 is locked out and Devices 0 and 1 can still interrupt. If Device 1 now generates an interrupt the servicing of Device 2 will be suspended in favor of Device 1 (see Figure 3.18c). On completion

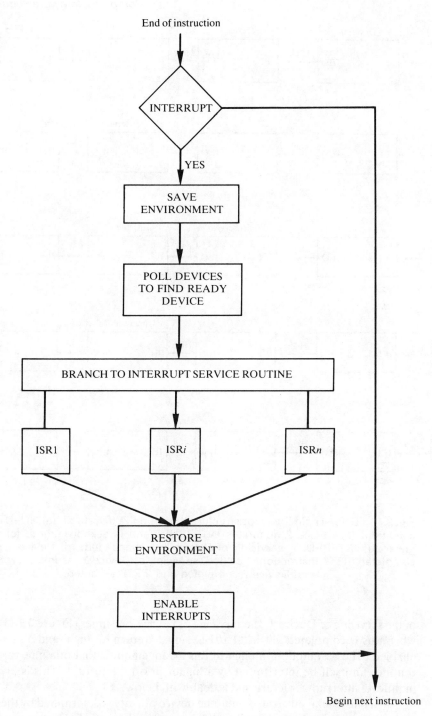

Fig. 3.17 Operations for software decoding of interrupts.

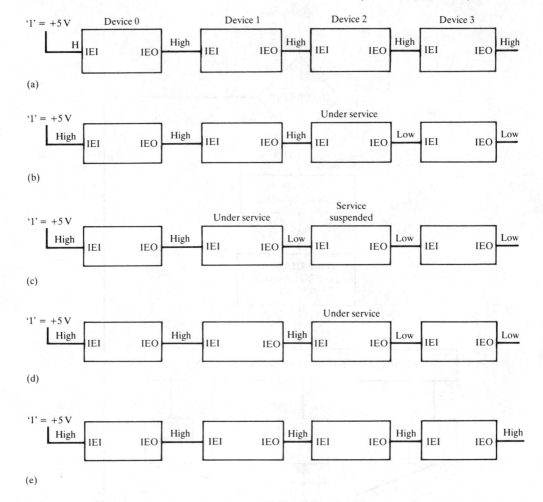

Fig. 3.18 Daisy-chain interrupt structure (based on Z80 system). (a) NO INTERRUPT condition) (b) Device 2 generates INTERRUPT and is acknowledged; (c) Device 1 generates INTERRUPT, servicing of Device 2 is suspended; (d) Device 1 servicing completed 'RETI' instruction executed, servicing of Device 2 resumed; (e) Servicing of Device 2 completed and 'RETI' executed.

of the servicing of Device 1, the servicing of Device 2 resumes (Figure 3.18d). Finally when this is completed all the IEI/IEO signals return to 'high' and all devices are enabled. The assumption implicit in this arrangement is that one interrupt service routine can itself be interrupted by a higher priority interrupt. This is known as a multilevel interrupt structure and is dealt with below.

In a daisy-chained arrangement the device priority is determined by the position of the device in the chain and cannot be changed by the software. Great care has to be given to timing considerations in daisy-chained arrangements; in particular, care has

Data Transfer Techniques: Interrupts 77

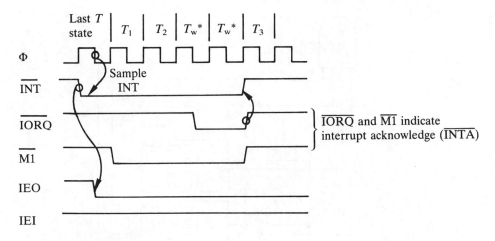

*Inserted by Z80-CPU

Fig. 3.19 Z80 interrupt response timing diagram.

to be taken to allow the IEI signal time to propagate along the chain. Figure 3.19 shows the timing diagram for the Z80 interrupt response. The M1 pin on the CPU is used to indicate that the CPU is in the first cycle of a machine code instruction (normally a fetch from memory operation); it is also used to indicate that it is beginning the interrupt acknowledge cycle, since the interrupt acknowledge signal is both M1 and IORQ active. When M1 and IORQ are active the interrupting device must put its interrupt vector on the data bus.

When M1 is active the devices must not change their interrupt request status. The reason for this is to prevent a higher priority interrupt; i.e., a device at the top of the chain interfering with a request from a lower priority device during the acknowledgement sequence for the lower priority device. Consider the position in which the lower priority device generates an interrupt and the interrupt is being acknowedged, but between M1 becoming active and the end of IORQ being active a higher priority device generates an interrupt request. The higher priority device lowers its IEO signal; it will take a small, but finite, time for this signal to pass along the chain to stop the lower device from placing its interrupt response vector on the data bus. The consequence could be that two devices attempt to place interrupt response vectors on the data bus. The delay between the activation of M1 and IORQ is to allow time for the propagation of the IEO signal. In Figure 3.19 the timing diagram shows two wait states inserted in the timing cycle; these are inserted automatically by the CPU for an interrupt acknowledge sequence and provide sufficient time for the chaining of up to four Z80 PIO devices. To provide for a longer chain or for different types of devices it may be necessary to force additional wait states in the interrupt acknowledge sequence by the use of the 'wait' pin on the CPU. An alternative is to use a different form of chaining for the IEI/IEO signals which provides a 'look-ahead' arrangement as shown in Figure 3.20.

The determination of interrupt priority can be performed using priority encoder

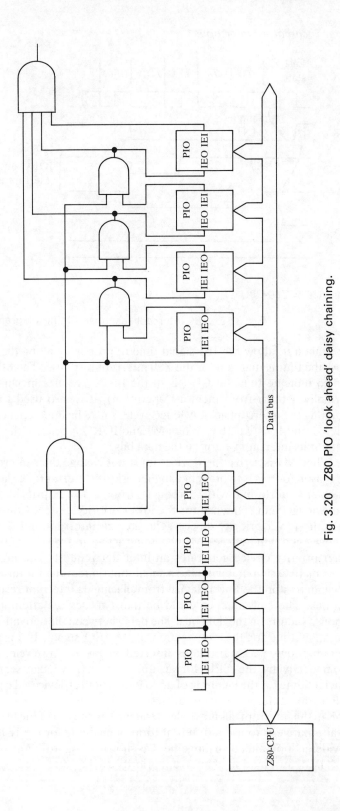

Fig. 3.20 Z80 PIO 'look ahead' daisy chaining.

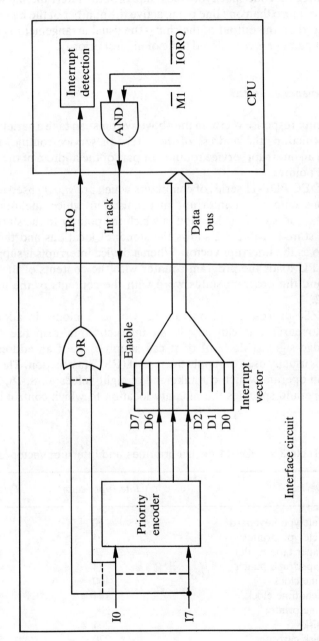

Fig. 3.21 Interrupt vectoring using priority encoding circuit.

circuits. A typical arrangement is shown in Figure 3.21. In this system an interrupt occurring on any line causes the interrupt common line to become active and also places a three bit code specifying the number of the interrupt line which is active. In the event of more than one line being active the number of the highest priority active line is placed on the output of the chip – the usual arrangement is that the line with the lowest number is considered to be of highest priority.

3.5.5 Interrupt response vector

The interrupt response vector in the above systems can take a variety of forms: it may be an instruction, the address of the interrupt service routine, the address of a pointer to an interrupt service routine, or part of the address of the interrupt service routine or pointer.

The DEC PDP-11 series of computers which are widely used in process control applications employ an interrupt mechanism in which the interrupting device supplies the address of the location in which the pointer to the start of the interrupt routine is stored. Table 3.2 shows the standard locations and these locations are referred to as the interrupt vector. When a device interrupts it supplies this address and the CPU loads the program counter with the contents of the interrupt vector location and the program status word with the contents of the location interrupt vector + 2.

The Z80 processor can operate in two other modes in addition to Mode 1 referred to earlier: it can supply an instruction in response to the interrupt acknowledge signal (Mode 0) or it can supply part of an address (Mode 2). In Mode 0 it is usually used to supply the restart (RST) instruction. The RST instruction requires an operand which can take one of eight values: 0, 8, 16, 24, 32, 40, 48 or 56. The operand specifies the memory location to which control is transferred on

Table 3.2 PDP-11 device priorities and interrupt vector locations

Device	Interrupt vector	Priority
Teletype keyboard	60	4
Teletype printer	64	4
Paper tape reader	70	4
Paper tape punch	74	4
Line clock	100	6
Real-time clock	104	6
Line printer	200	4
Disk	204	4
Disk cartridge	220	5
Magnetic tape	224	5

Data Transfer Techniques: Interrupts

receipt of the interrupt. One of these locations therefore forms the start of the interrupt service routine.

The Mode 0 method is easily implemented with a simple interface as is shown in Figure 3.22. (It should be noted that a priority encoding circuit would normally be used rather than a simple 8-line to 3-line encoder. Why?) The bit pattern for the instruction is:

$$11\,xxx\,111$$

where the bits xxx are used to indicate the appropriate response location as shown in Table 3.3. Hence by connecting the output bits from a priority encoder to the data lines 3, 4 and 5 with the other lines set to 1, the correct RST instruction will be generated.

Instead of responding with an RST instruction the device could respond with a 'call' instruction; however, in this case the external hardware required is more complicated because a 'call' instruction requires two further words of data to be placed on the data bus: the low byte and then the high byte of the address for the 'call'

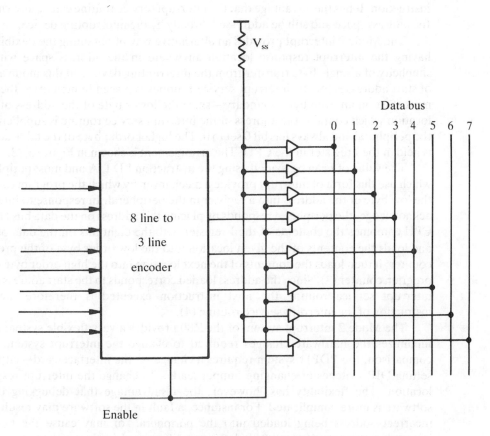

Fig. 3.22 Simple implementation of Mode 0 interrupt.

Table 3.3 Location of pointers for interrupt routines

xxx	Location
000	0
001	8
010	16
011	24
100	32
101	40
110	48
111	56

instruction. It has the advantage that the interrupt service routine can be anywhere in the address space and still be addressed directly by the interrupting device.

The Mode 2 interrupt provides an alternative way of obtaining the flexibility of having the interrupt response location anywhere in the address space with the simplicity of a single byte transfer from the interrupting device. In this mode a table of start addresses for the interrupt service routines is placed in memory – the table must start on an even byte boundary – since the lower byte of the address of table location which contains the address of the interrupt service routine is supplied from the peripheral and always has bit 0 set to 0. The higher order byte of the table address is held in the I register in the CPU. The arrangement is shown in Figure 3.23.

The value of I can be loaded using the instruction LD I, A and most peripherals which use this form of interrupt provide a mechanism by which the program can load the low byte of the address into a register in the peripheral. In response to interrupt acknowledge, the peripheral responds by placing this address on the data bus (1); the CPU combines the contents of the I register with the eight bits on the data bus (2) and loads the contents of the given location into the low order byte of the program counter; it then loads the contents of the next location into the high order byte of the program counter (3). Since the address loaded corresponds to the start address of the interrupt service routine the next instruction executed is therefore the first instruction of the interrupt service routine (4).

The Mode 2 interrupt system of the Z80 provides a very flexible system and it minimizes the hardware changes required to change the interrupt system. As a comparison, the PDP11 system requires a change on the interface card – either by setting DIP switches or changing jumper leads – to change the interrupt response location. The flexibility has, however, the disadvantage that debugging of the software is more complicated. For instance, a fault in the software may result in an incorrect address being loaded into the peripheral, or may cause the table of interrupt service routines to be loaded into a different place than expected.

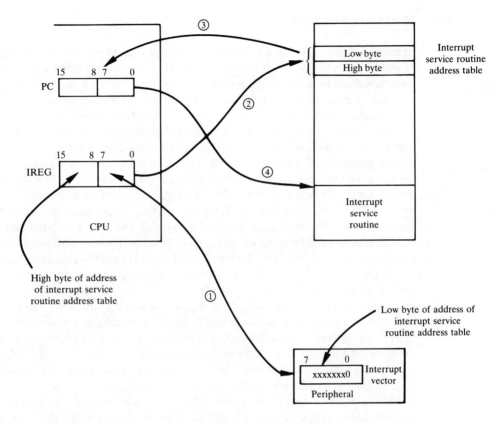

Fig. 3.23 Sequence of operations for Z80 Mode 2 interrupt structure.

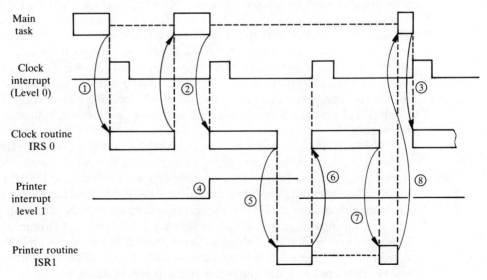

Fig. 3.24 Multilevel interrupts.

3.5.6 Multilevel interrupts

In most real-time systems a single interrupt level is unacceptable: the whole purpose of interrupts is to get a fast response and this would be prevented if a low priority interrupt could lock out a high priority one. A typical picture of multilevel interrupts is shown in Figure 3.24. An application program is interrupted at regular intervals by the clock interrupt which is the highest priority interrupt (level 0). When the interrupt occurs control is passed to the clock interrupt service routine (ISR 0): transfers 1, 2 and 3 in Figure 3.24. During the servicing of the clock interrupt, the printer generates an interrupt request (4), since the printer is of lower priority than the clock the interrupt cannot be dealt with until the clock routine ISR 0 has finished. When this occurs, control, instead of returning to the main program, passes to the printer service routine ISR 1 (5). The printer service routine cannot complete before the next clock interrupt so it is suspended (6) while the next interrupt from the clock is dealt with. At the termination of the clock routine return is made to the printer (7) and finally when the printer ISR finishes, a return is made to the main program (8).

It should be obvious that the ability to interrupt an interrupt service routine should be restricted to interrupts which are of higher priority than the routine executing. In order to do this there has to be some facility for masking out (or inhibiting) interrupts of lower priority. Masking is achieved automatically in the daisy chain system, since a device which wishes to interrupt lowers its IEO line which prevents all lower priority devices from responding to the interrupt acknowledge signal. In the daisy-chain system, however, the device must receive a signal from the CPU on return from the interrupt service routine in order that it can set the IEO line high and hence permit access to the system by lower priority devices. For example, all Z80 support chips recognize the RETI instruction and use it to reset the IEO line high.

An alternative scheme used is to have a mask register which can be loaded from software and used to inhibit the lower priority interrupt lines. In Figure 3.25 a system is shown in use with a priority encoder; the software sequence is outlined in Figure 3.26. Note that with the mask system it is possible to mask out any interrupt, not just ones with lower priority; this can have advantages if for example a high priority alarm interrupt is continually being generated because of a fault on the plant – once the fault condition has been recognized it is desirable to mask out the interrupt to avoid the computer spending all its time simply servicing the interrupt. The ability to mask out selected levels provides a means of re-allocating priorities from the software.

The functions to be performed by a typical interrupt service routine are shown in Figure 3.26. As in the software decoded routine the first requirement is to save the working environment; the current mask register must also be saved and then the new mask register sent out. The interrupts can now be enabled and the actual servicing of the interrupt commenced. When the servicing is completed the interrupts should be disabled, the previous mask register restored and the working environment restored. The interrupts can now be enabled and a return from interrupt executed. It should be noted that some computer systems automatically disable all lower priority interrupts and hence the need to save and restore mask registers is avoided.

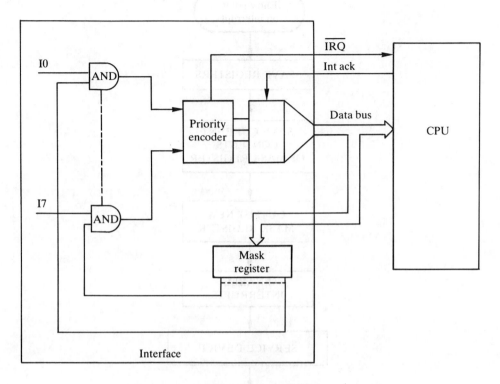

Fig. 3.25 Interrupt masking.

3.6 COMPARISON OF DATA TRANSFER TECHNIQUES

The use of polling with either busy-wait or periodic checks on device status provides the simplest method in terms of the programming requirements and in the testing of programs. The use of interrupts results in software which is much less structured than a program with explicit transfers of control: in an interrupt-driven program there are potential transfers of control at every point in the program.

Interrupt-driven systems are much more difficult to debug since many of the errors may be time-dependent. A simple rule is to check the interrupt part of the program if irregular errors are occurring. The generation of appropriate test routines for interrupt systems is difficult; for proper testing it is necessary to generate random interrupt patterns and to carry out detailed analysis of the results.

At high data-transfer rates the use of interrupts is inefficient because of the overheads involved in the interrupt service routine – saving and restoring the environment – hence polling is often used. An alternative for high rates of transfer is to substitute hardware for software control and use direct memory access techniques.

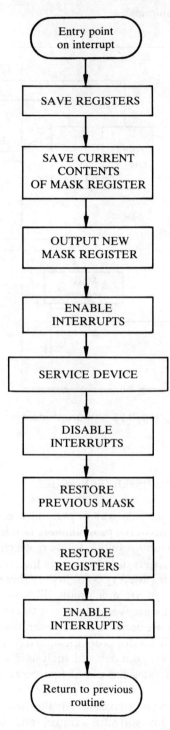

Fig. 3.26 Function performed by interrupt servicing routine.

3.6.1 Direct memory access

Three methods are normally used: burst mode, distributed mode and cycle stealing. In burst mode, the DMA controller takes over the data highways of the computer and locks out the CPU for the period of time necessary to transfer, say, 256 bytes to or from fast memory to backing memory. The use of burst mode can seriously affect the response time of a real-time system to an external event and because of this may not be acceptable.

In distributed mode the DMA controller takes occasional machine cycles from the CPU's control and uses each cycle to transfer a byte of information to or from fast memory to the backing memory. In a non-real-time system the loss of these machine cycles to the CPU is not noticeable; however, in a real-time system, if software timing loops are used then the loss of machine cycles will affect the time taken to complete the loop. Note that a software timing loop relies on the time taken to complete a specified number of machine cycles remaining constant; if a DMA controller is using some of the cycles, this will not be known to the program which will still cycle through the same number of instructions, but instead of this involving, say, 100 machine cycles, it may take 200.

The cycle-stealing method transfers data only during cycles when the CPU is not using the data bus. Therefore the program proceeds at the normal rate completely unaffected by DMA data transfers. This is, however, the slowest method of transfer to backing store.

3.7 COMMUNICATIONS

The use of distributed computer systems implies the need for communication: this covers communication between instruments on the plant and the low-level computers (see Figure 3.27); between the Level 1 and Level 2 computers and between the Level 2 and the higher level computers. At the plant level the communications systems used typically involve parallel analog and digital signal transmission techniques. At the higher levels it is more usual to use serial communication methods.

As the distance between the source and receiver increases it becomes more difficult, when using analog techniques, to obtain a high signal-to-noise ratio; this is particularly so in an industrial environment where there may be numerous sources of interference. Analog systems are therefore generally limited to short distances. The use of parallel digital transmission provides high data transfer rates but is expensive in terms of cabling and interface circuitry and again is normally only used over short distances (or when very high rates of transfer are required).

Serial communication techniques can be characterized in several ways:

1. Mode   (a) asynchronous
                   (b) synchronous

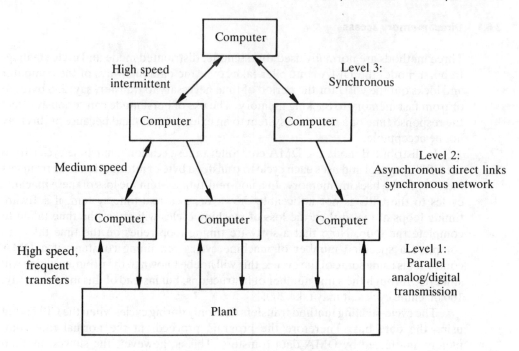

Fig. 3.27 Data transmission links.

2. Quantity   (a) character-by-character
                   (b) block
3. Distance   (a) local
                 (b) remote i.e. wide area
4. Code   (a) ASCII
                 (b) other

3.7.1 Asynchronous and synchronous transmission techniques

Asynchronous transmission implies that both the transmitter and receiver circuits use their own local clock signals to gate data on and off the data transmission line. In order that the data can be interpreted unambiguously there must be some agreement between the transmitter and receiver clock signals. This agreement is forced by the transmitter periodically sending synchronization information down the transmission line.

The most common form of asynchronous transmission is the character-by-character system which is frequently used for connecting terminals to computer equipment and was introduced for the transmission of information over telegraph lines. It is sometimes called the stop-start system. In this system each character which is transmitted is preceded by a 'start' bit and followed by one or two 'stop' bits (see Figure 3.28). The start bit is used by the receiver to synchronize its clock with the incoming data; the signal must remain synchronized for the time taken to receive

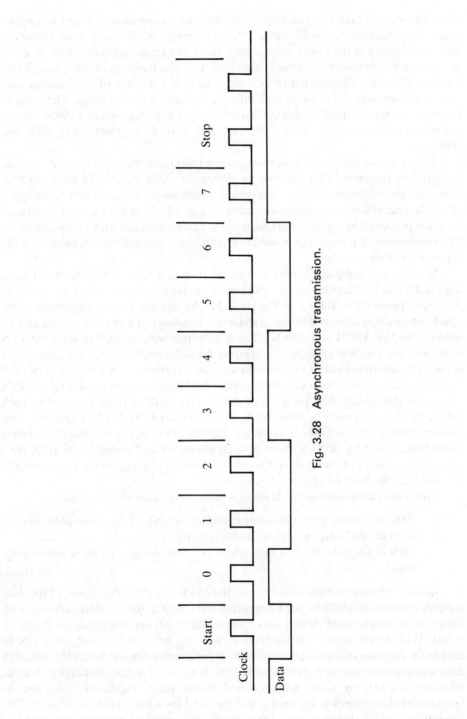

Fig. 3.28 Asynchronous transmission.

the following eight data bits and two stop bits. The transmission is thus bit synchronous, but character asynchronous. The advantage of the stop-start system is that, particularly at the lower transmission rates, the frequencies of the clock signal generators do not have to be closely matched. The disadvantage of the system is that for each character transmitted (7 bits) three or four extra bits of information have also to be transmitted; i.e, the overall information ratio is not very high. The range of transmission speeds used for this system is from 75 bits per second up to 9600 bits per second (the standard speeds are 75, 110, 300, 600, 1200, 1800, 2400, 4800 and 9600).

To overcome the problem of transmitting redundant bits, synchronous systems designed to transmit large volumes of data over short periods of time, such as computer-to-computer systems, use block synchronous transmission techniques. Here, the characters are grouped into records, e.g., blocks of 80 characters, and each record is preceded by a synchronization signal and terminated with a stop sequence. The synchronization sequence is used to enable the receiver to synchronize with the transmitter clock.

In order to establish effective communication it is necessary to transmit more than just a synchronization signal – the additional information is called the *protocol*. A simple protocol is shown in Figure 3.29. At the start of a transmission, bit synchronization is achieved by the transmitter sending out a sequence of 0s and 1s, followed by the ASCII code 'SYN'. The transmitter will continue to send the SYN code until the receiver responds by sending back the code 'ACK' or a preset time elapses (device time out); if time out occurs, the transmitter sends the bit pattern of 0s and 1s again. Once contact has been established the transmitter will send out SYN characters during any idle period and the receiver will respond by sending back ACK; the line will only be completely idle when an EOT (end of transmission) character has been sent by the transmitter. The text is broken up into blocks and each block is preceded by an STX (start of text) character and ended by an ETX (end of text) character. Following the ETX will be an integrity check on the data; typically this will take the form of a parity check.

There are two main standards for synchronous transmission systems:

1. BISYNC (binary synchronous communication) This is the older system used in IBM equipment and is obsolescent.
2. HDLC (high-level data link control) This is used in most new equipment.

In synchronous transmission systems the clock signal for the timing of the data transfer is provided solely by the transmitter and is sent to the receiver even when no data is being transmitted. When data is transmitted it is superimposed on the clock signal. With synchronous transmission there is no need to transmit extra bits to enable the receiver clock to synchronize with the transmitter and hence the effective data transmission rate for a given speed of line is higher. The disadvantage is that the interface circuitry is more complex and hence more expensive. The use of synchronous transmission does not avoid the need for a transmission protocol. The advantage of block transmission is that a much higher ratio of data bits to control bits can be obtained.

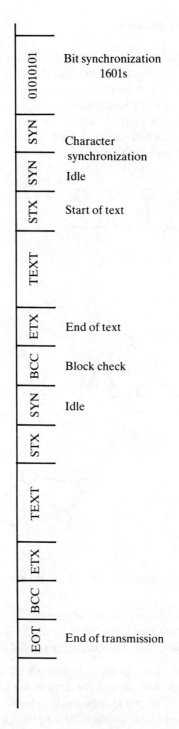

Fig. 3.29 Synchronous transmission.

3.7.2 Local and wide area networks

Wide area networks have existed for many years and they operate over a very wide geographical area (many are international networks) at moderate speeds. The local area network (LAN) is a more recent development and it is having a considerable impact on the design of process control equipment. LANs make use of a wide range of transmission media such as twisted pair, co-axial cable, and fiber optics; they operate at a range of transmission speeds (up to 240 Mbits/s) and use a range of different protocols and topologies.

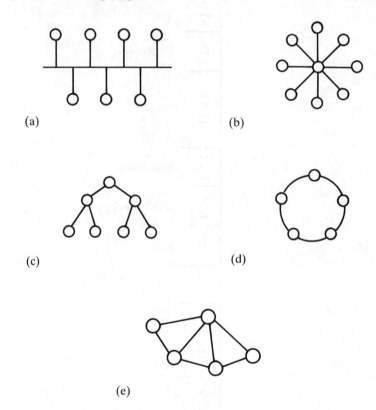

Fig. 3.30 LAN topologies. (a) Data bus; (b) star; (c) hierarchy; (d) ring; (e) mesh.

Typical topologies are shown in Figure 3.30. For computer control applications no one topology represents the best solution: the particular application will govern the most appropriate one. The characteristics of each are briefly outlined below.

1. *Data bus* This is the simplest of all the LAN topologies. The bus is normally passive and all the devices are simply plugged into the transmitting medium. The bus is inherently reliable because of its passive nature but there may be limitation on the length of a bus in that any transmitting device

connected has to be able to transmit for the full length of the bus. A packet of data placed on the bus is available to all devices, i.e., it is a broadcast system. The major weakness of the bus system is the mechanism of data access, i.e. when a device can put information onto the bus.

2. *Star* The star network is not very widely used; it depends on a central switching node to which all other nodes are connected by a bidirectional link. Data sent to the central switch can be forwarded either in the broadcast mode, i.e., to all other nodes, or only to a specified node. The computer controlled PABXs (private automatic branch exchanges) used in many businesses operate in the star mode – all the telephone lines connect to the central unit – the forwarding system used is to a specified node. A weakness of the star topology is that the central node is a critical component in the system; if it fails, the whole system fails.

3. *Hierarchy* The system has many of the characteristics of the star, but instead of one central switching node, many of the nodes have to act as switches. It can frequently closely reflect the actual structure of the application. The addition of new nodes to a hierarchy can be difficult.

4. *Ring* This is probably the most popular method. The ring is typically an active transmission system, i.e., the ring itself contains regeneration circuits which amplify the signals. The information placed on a ring network continues to circulate until a device removes it from the ring; in some systems the originating device removes the data from the ring. The information is broadcast in the sense that it is available to all devices connected to the ring.

5. *Mesh* The mesh topology allows for random interconnection between the various nodes. It provides a means by which alternate routes between nodes can be found and hence has built into it a form of redundancy. A problem which can arise with the mesh is that there can be a delay between the sending and receiving of the message because of the number of nodes through which the message has had to pass.

Information is transmitted in the form of 'packets' which may be of fixed or variable length. Early systems used character-oriented packets similar to that illustrated in Figure 3.29. The prevailing standard is now the HDLC protocol (referred to in the previous section) and the format of the packet is shown in Figure 3.31.

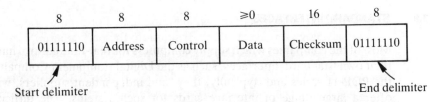

Fig. 3.31 Format of HDLC packet.

The *access* mechanisms used to ensure that only one node on the network is attempting to transmit at any one time divide into two main types:

- *synchronous* — token passing; message slots
- *asynchronous* — carrier sense multiple access/collision detection (CSMA/CD)

Ring-based LANs normally use synchronous techniques. The packets of data circulate in one direction round the ring, and in the token passing system attached to one, and only one, packet is a token. (If the network is idle a packet containing just the token is circulated.) Each node reads the packet into a buffer and checks the message. There are three actions which can then be taken.

1. If the message is for that node it is read, marked as accepted and replaced on the ring – it will be removed from the ring when it reaches the originating node.
2. If the message is not for that node it is simply replaced on the ring.
3. If the packet contains the token and the node wishes to transmit a message, the token is removed from the packet which is then passed on. The node then transmits its own message adding the token to the end of the message.

An alternative to the token passing method is the message slot. A sequence of bits is used to mark a slot and the slots circulate around the ring. If a node detects an empty slot it may insert a message in that slot.

Bus LANs may use token passing (the token is passed from node to node in some predetermined manner) or message slots but asynchronous methods are more common. In the asynchronous systems a node may attempt to transmit at any time. The node listens to the bus and if it is idle begins to transmit. Because of the distances between nodes and the time taken to transmit a message, two (or more) nodes may be transmitting simultaneously. If this happens a collision is said to have occurred; bus systems must therefore have some means of detecting a collision. If a collision is detected the nodes attempting to transmit execute the random delay and then retry.

Several of the major process companies have developed distributed control systems based on the use of LAN technology (mostly rings). These allow a wide range of devices – from individual instruments to large computers – to be connected to a common network.

3.8 STANDARD INTERFACES

Most of the companies which supply computers for real-time control have developed their own 'standard' interfaces, such as the Digital Equipment Company's Q-bus for the PDP-11 series and, typically, they, and independent suppliers, will be able to offer a large range of interface cards for such systems. The difficulty with the standards supported by particular manufacturers is that they are not compatible with each other, hence a change of computer necessitates a redesign of the interface.

An early attempt to produce an independent standard was made by the British Standards Institute (BS 4421, 1969). Unfortunately the standard is limited to the concept of how the devices should interconnect and the standard does not define the hardware. It is not widely used and has been overtaken by more recent developments.

An interface which was originally designed for use in atomic energy research laboratories – the computer automated measurement and control (CAMAC) system – has been widely adopted in laboratories, the nuclear industry and some other industries. There are over 1000 different modules, supported by about 50 manufacturers in eight countries, available for the system. There are also FORTRAN libraries which provide software to support a wide range of the interface modules. One of the attractions of the system is that the CAMAC data highway connects to the computer by a special card: to change to a different computer only requires that the one card be changed.

A general purpose interface bus (GPIB) was developed by the Hewlett Packard Company in the early 1970s for connecting laboratory instruments to a computer. The system was adopted by the IEEE and standardized as the IEEE488 bus system. The bus can connect up to a maximum of 15 devices and is only suited to laboratory or small, simple control applications.

EXERCISES

3.1 Why is memory protection important in real-time systems? What methods can be used to provide memory protection?

3.2 A large valve controlling the flow of steam is operated by a dc motor. The motor controller has two inputs:
(a) on/off control, $0\,V$ = off, $5\,V$ = on; and
(b) direction $0\,V$ = clockwise, $5\,V$ = anti-clockwise;
and two outputs:
(a) fully open = $5\,V$
(b) fully closed = $5\,V$.
Show how this valve could be interfaced to a computer controlling the process.

3.3 A turbine flowmeter generates pulses proportional to the flowrate of a liquid. What methods can be used to interface the device to a computer?

3.4 There are a number of different types of analog-to-digital converters. List them and discuss typical applications for each type (see, e.g., Woolvet or Barney).

3.5 The clock on a computer system generates an interrupt every 20 ms. Draw a flowchart for the interrupt service routine. The routine has to keep a 24-hour clock in hours, minutes and seconds.

3.6 Twenty analog signals from a plant have to be processed (sampled and digitized) every 1 second. The analog-to-digital converter and multiplexer which is available can operate in two modes: automatic scan and computer-controlled scan. In the automatic scan mode, on receipt of a 'start' signal the converter cycles through each channel in turn. The data corresponding to the channel sampled is available for 0.9 ms. The signal 'not-ready' is asserted during the conversion period and this indicates that the data is changing and should not be read by the computer. The timing is shown in Figure 3.32a. In Mode 2 under computer-controlled scanning, the converter holds the data for each channel sampled until it receives a command from the computer to start the sampling of the next channel. To speed up the operation the multiplexer is switched to the next channel once the current channel has been sampled and before the computer reads the data for the current channel. The converter can be reset to start from Channel 1 by asserting a signal reset. The timing of this mode of operation is shown in Figure 3.32b. Consider the ways in which (a) polling and (b) interrupt methods can be used to interface the converter to a computer. Discuss in detail the advantages and disadvantages of each method.

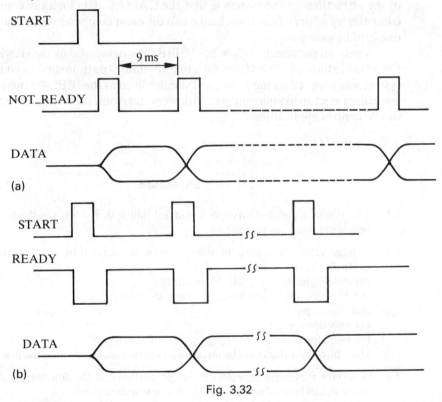

Fig. 3.32

3.7 We will assume that the simple heat process described in Chapter 1 has in, addition to a temperature sensor and heat controller, some additional logic signals and control switches. These are:

Plant controls
  heater on/off;
  blower on/off; and
  power on/off.
Plant signals
  over-temperature alarm; and
  blower failure alarm.

The start up sequence for the unit is:

1. Turn power on.
2. Turn blower on.
3. Wait five seconds.
4. Turn heater on.
5. DDC control action begins.

If at any time the over-temperature alarm becomes true, i.e., the signal level is set to logic 'high', the heater must be turned off but the blower kept running. If the blower failure alarm is detected, both the blower and the heater must be switched off.

Draw a flowchart to show the sequence of operations to be carried out (a) for start up, and (b) in the event of failure.

3.8 The hot-air blower system described in Chapter 1 uses interrupts to indicate: power failure, printer ready, air temperature too high, VDU display ready, blower failure, clock signal and key pressed on the keyboard. Draw up a list of the priority order for the interrupts and explain the reasons for your choice of priority. Should any of the interrupts be connected to a non-maskable interrupt?

REFERENCES AND BIBLIOGRAPHY

A large number of books deal with microprocessors, digital interfaces, instrumentation, etc. A small selection of them is given below. Good introductory books on digital logic systems are Mano (1979) and Tocci (1980). For a general introduction to small computer systems see Ffynlo-Craine (1985), Lippiatt (1978) or Mellichamp (1983). For transducers and interfacing see Andrews (1982), Barney (1985), Johnson (1984), Mellichamp (1983), Sargeant (1981), Stone (1983) and Woolvet (1977). Introductions on local area networking are to be found in Hopper (1986) and Sloman (1982).

ANDREWS, M. (1982), *Programming Microprocessor Interfaces for Control and Instrumentation*, Prentice Hall

BARNEY, G.C. (1985), *Intelligent Instrumentation*, Prentice Hall

BENNETT, S. and LINKENS, D.A. (eds.) (1982), *Computer Control of Industrial Processes*, Peter Peregrinus, Stevenage

CASSELL, D.A. (1982), *Microcomputers and Modern Control Engineering*, Prentice Hall

FFYNLO-CRAINE, J. and MARTIN, G.R. (1985), *Microcomputers in Engineering and Science*, Addison Wesley

GORSLINE, G.W. (1986 2nd edition), *Computer Organisation*, Prentice Hall

HOPPER, A., TEMPLE, S. and WILLIAMSON, R. (1986), *Local Area Network Design*, Addison Wesley

LIPPIATT, A.G. (1978), *The Architecture of Small Computer Systems*, Prentice Hall

MANO, M.M. (1979), *Digital Logic and Computer Design*, Prentice Hall

MELLICHAMP, D.A. (ed.) (1983), *Real-time Computing with Applications to Data Acquisition and Control*, Van Nostrand, New York

SARGEANT, M. and SHOEMAKER, R.L. (1981), *Interfacing Microcomputers to the Real World*, Addison Wesley

SLOMAN, M.S. (1982), 'Communications for distributed control' in Bennett, S. and Linkens, D.A. (eds.) (1982)
SLOMAN, M.S. and KRAMER, J. (1986), *Distributed Systems and Computer Networks*, Prentice-Hall
STONE, H.S. (1983), *Microcomputing Interfacing*, Addison Wesley
TOCCI, R.J. (1980), *Digital Systems: Principles and Applications*, Prentice Hall
WOOLVET, G.A. (1977), *Transducers in Digital Systems*, Peter Peregrinus, Stevenage

4

DDC Control Algorithms and their Implementation

4.1 INTRODUCTION

In this chapter we will consider DDC control algorithms, various methods used to implement them and some of the practical problems which occur in digital implementations. A typical control system is shown in continuous form in Figure 4.1a and discrete form in Figure 4.1b. Two main forms of $G_c(s)$ will be considered:

- three-term control; and
- controllers designed using plant models.

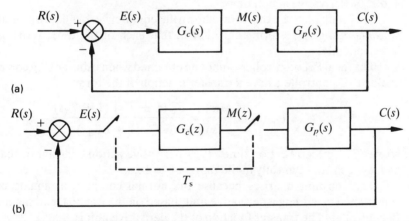

Fig. 4.1 General form of a control system. (a) Continuous form; (b) discrete form.

The three-term (PID) control will be considered because it is still very widely used in industry. The reasons for its widespread use are that it requires very little knowledge of the plant dynamics and the methods of determining the controller parameters are well known and understood [Auslander 1981, Ahson 1983, Cohen 1953, Leigh 1985, Smith 1972, Stephanopoulos 1985, Takahashi 1970, Ziegler 1942]. If knowledge of the plant model is available a wide variety of techniques can be used. The design can be carried out using continuous system design methods to find $G_c(s)$ followed by discretization of $G_c(s)$ to give $G_c(z)$. A number of methods of mapping $G_c(s)$ to $G_c(z)$ are available (see section 4.9) but none gives a controller $G_c(z)$ with exactly the same characteristics of $G_c(s)$. Care must be used since, for example, discretization using first order finite differences can easily result in changing a stable continuous-time system into an unstable discrete-time system. An alternative is to design $G_c(z)$ on the basis of a discrete-time model of the plant, $G_p(z)$, obtained either by discretization of the continuous-time model $G_p(s)$ or by determining $G_p(z)$ directly.

In both cases we are concerned with how the controller is programmed, not with how it is designed (information on design techniques can be found in Franklin 1980, Iserman 1981, Katz 1981, Kuo 1980, Leigh 1985, and Smith 1972).

4.2   THE PID CONTROL ALGORITHM: THE BASIC ALGORITHM

The PID controller can be expressed in the form

$$G_c(s) = \frac{M(s)}{E(s)} = K_c \left(1 + \frac{1}{T_i s} T_d s\right) \qquad (4.1)$$

where k_c = controller gain (i.e. 1/proportional band), T_i = integral action time, and T_d = derivative action time.

The controller in the form shown in equation 4.1 is known as the *ideal* or *non-interacting* three-term controller in that each of the three terms is evaluated independently.

The ideal PID controller cannot be obtained when using analog components and analog PID controllers have a transfer function of the form

$$G_c(s) = \frac{M(s)}{E(s)} = \frac{K_c (1 + T_i s)(1 + T_d s)}{T_i s (1 + \alpha T_d s)} \qquad (4.2)$$

where T_i = integral action time, T_d = derivative action constant of real controller and α = constant typically in the range 1/6 to 1/20.

The constant α arises because the normal commercial analog controller is constructed by adding a derivative unit, based on the use of a lead-lag network, to a PI controller. The transfer function of the derivative unit is

$$\frac{D(s)}{E(s)} = \frac{(T_d s + 1)}{(\alpha T_d s + 1)}$$

The ideal PID controller can be implemented on a digital computer but it is usual to incorporate a filter of some form to provide for high frequency roll-off for the derivative action term. The desired roll-off can be achieved by (a) filtering the input to the ideal controller, (b) filtering the output, or (c) using a form of controller which incorporates a roll-off component, i.e., implementing a controller which incorporates a lead-lag component rather than pure derivative action. In the next section a method of implementing the ideal controller is described; it is assumed that filtering of either the input or the output is used.

4.3 IMPLEMENTING THE IDEAL PID CONTROLLER

Equation 4.1 can be expressed in the time domain in the following form

$$m(t) = K_c T_d \frac{de(t)}{dt} + K_c e(t) + \frac{K_c}{T_i} \int e(t)dt \qquad (4.3)$$

In order to derive a control algorithm, equation 4.3 has to be converted to discrete form. One method of doing this is to use first order finite differences; we can write

$$\frac{df}{dt}\bigg|_k = \frac{f_k - f_{k-1}}{\Delta t}, \quad \int e(t)dt = \sum_{k=0}^{n} e_k \cdot \Delta t \qquad (4.4)$$

and hence equation 4.3 becomes

$$m_n = K_c \left[T_d \frac{(e_n - e_{n-1})}{\Delta t} + e_n + \frac{1}{T_i} \sum_{k=0}^{n} e_k \cdot \Delta t \right] \qquad (4.5)$$

By introducing new parameters as follows:

$$K_p = K_c$$
$$K_i = K_c * T_s / T_i$$
$$K_d = K_c * T_d / T_s$$

where $T_s = \Delta t =$ the sampling interval, equation 4.5 can be expressed as an algorithm of the form

$$s_n = s_{n-1} + e_n$$
$$m_n = K_p * e_n + K_i * s_n + K_d (e_n - e_{n-1}) \qquad (4.6)$$

where $s_n =$ the sum of the errors.

A simple PID controller using the alogrithm given in equation 4.6 can be programmed in Pascal as follows:

* * *

Example 4.1 PID controller written in Pascal

```
PROGRAM PIDController;
CONST
     kpvalue = 1.0;
     kivalue = 0.8;
     kdvalue = 0.3;
VAR
     s,kp,ki,ke,en,enold,mn  : REAL;
     stop                    : BOOLEAN;
FUNCTION ADC:REAL; EXTERNAL;
PROCEDURE DAC(VAR mn:REAL); EXTERNAL;
BEGIN (* main program *)
     stop := FALSE;
     s := 0.0;
     kp := kpvalue;
     ki := kivalue;
     kd := kdvalue;
```

```
    enold := ADC;
(* control loop *)
REPEAT
    en := ADC; (* ADC returns value of error en *)
    s := s+en; (* integral summation *)
    mn := kp*en + ki*s + kd*(en - enold);
    DAC(mn);
  enold := en;
UNTIL stop;
END.
```

* * *

The same program can be written in FORTRAN.

* * *

Example 4.2   PID controller written in FORTRAN

```
PROGRAM PIDCONTROLLER;
C    DECLARE VARIABLES
     REAL EN, ENOLD, MN, S
     REAL KI, KP, KD
     REAL KIVALUE, KPVALUE, KDVALUE
     PARAMETER (KPVALUE=1.0, KIVALUE=0.8, KDVALUE=0.3)
     LOGICAL FINISHED
C DECLARE EXTERNAL SUBROUTINES
     EXTERNAL ADC, DAC
C INITIALIZE DATA
     FINISHED = .FALSE.
     S=0.0
     KP=KPVALUE
     KI=KIVALUE
     KD=KDVALUE
     CALL ADC(ENOLD)
C
C   MAIN CONTROL LOOP
C
100  CALL ADC(EN)
     S=S+EN
     MN=KP*EN + KI*S +KD*(EN - ENOLD)
     CALL DAC(MN)
     ENOLD = EN
     IF (.NOT.FINISHED) GOTO 100
END
```

* * *

The programs shown in Examples 4.1 and 4.2 are both incorrect and impracticable. They are incorrect because they do not take into account the need to synchronize the calculation of the control variable with 'real' time. As the programs are written, the sampling time T_s is dependent on the speed of operation of the computer on which the program is run. For correct operation some means of fixing the time interval is required.

The programs are impractical because the form of the PID algorithm used does not take into account actuator limiting, which can cause integral wind-up; or measurement and process noise; or the need for smooth transfer from manual to automatic control. The program also has the controller parameters built in as program constants, hence modification of the controller settings requires recompilation of the program.

4.4 TIMING

An essential feature of real-time programs is that they have to run continuously so that the natural structural element is the infinite loop (this is represented by the programming construct LOOP...END); hence the general form of a control program will be:

* * *

Example 4.3 General form of a control program

```
PROGRAM RealTimeControl;
(* declarations *)
BEGIN
   LOOP
     ControlTask;
   END (* loop *);
END RealTimeControl.
```

* * *

As we have already noted the actions of ControlTask have to be synchronized with the requirements of the process being controlled. In many cases this requirement will be that the control action is performed at a specified sampling rate. The procedure ControlTask must therefore contain some means by which synchronization can be achieved.

The methods available are:

- polling;
- external interrupt;
- ballast coding; and
- real-time clock signals.

The first two methods, polling and external interrupt, rely on the process sending a signal when it is ready for a control action to take place. This signal must be sent at the sampling interval T_s chosen when starting the controller since, as can be seen from equation 4.5, the algorithm is correct only for a particular sampling rate. The difference between the two methods is that in polling the control computer repeatedly reads a value – normally a logical signal – whereas with an external interrupt the computer can be performing other computations, the action of the interrupt is to tell the computer to suspend whatever it is doing and carry out the control task.

An example of PROCEDURE ControlTask using polling is as follows.

* * *

Example 4.4 Use of polling for timing

```
PROCEDURE ControlTask;
BEGIN
  LOOP
    WHILE NOT(Digin(SampleTime)) DO
    (* wait until time *)
    END (* while *);
(* control action *)
  END (* loop *);
END ControlTask;
```

Example 4.5 Use of real-time clock for timing

```
MODULE PIDController;
CONST
        kpvalue = 1.0;
        kivalue = 0.8;
        kdvalue = 0.3;
sampleInterval = 0.01
VAR
    s,kp,ki,ke,en,enold,mn : REAL;
    time, nextSampleTime : REAL;
                     stop : BOOLEAN;
FROM PlantIO IMPORT
            ADC, GetTime, DAC;
BEGIN (* main program*)
  stop := FALSE;
    S := 0.0;
   kp := kpvalue;
   ki := kivalue;
   kd := kdvalue;
 enold := ADC;
  time := GetTime;
nextSampleInterval := time + sampleInterval;
(* control loop *)
  LOOP
    WHILE  time < nextSampleTime DO
        time := GetTime
    END (* while *);
     en := ADC; (* ADC returns value of error en *)
      s := s+en; (* integral summation *)
     mn := kp*en + ki*s + kd*(en - enold);
    DAC(mn);
  enold := en;
  nextSampleTime := time + sampleInterval;
    END (* loop *);
  END PIDController.
```

* * *

Alternative Forms of the PID Algorithm

In Example 4.4 it is assumed that a Boolean function, Digin(line), is available which reads the appropriate logical signal, in this case SampleTime, from the process interface.

The polling method is restricted to small dedicated systems, but has the advantage of simplicity and ease of programming. An alternative for simple, dedicated, control systems is the use of ballast coding [Hine 1979]. In general, the time taken to execute the code for a digital controller will depend on the values of the data. Even for

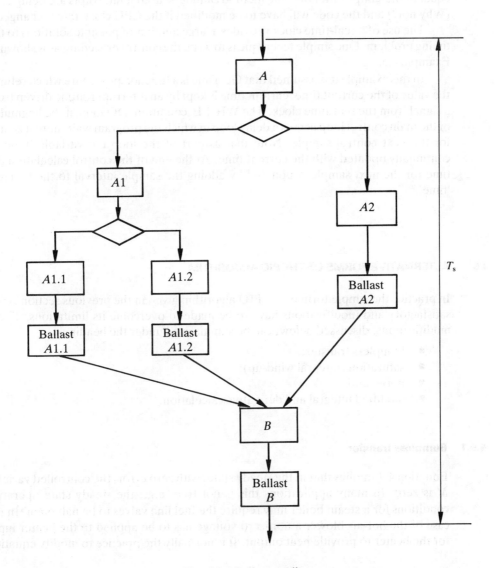

Fig. 4.2 Ballast coding.

the code shown in Examples 4.1 and 4.2, which do not contain any branches, the time will vary since the underlying arithmetic operations will have varying execution times, dependent upon the sign and actual values of the variables. For a simple controller programmed in assembler it is possible to construct a diagram showing the various possible paths through the code. An example is given in Figure 4.2. For each path (e.g., $A, A1, A1.1, B$) the computational time for that particular path is calculated (or measured) and code is added to each so as to make the computational time for each path equal. Further ballast code can be added to make the total computational time equal to the sample interval. The method cannot be used if interrupts are being used (Why not?) and the code will have to be modified if the CPU clock rate is changed.

The use of a real-time clock provides a large number of possible solutions to the timing problem. One simple technique is to write the control procedure as is shown in Example 4.5.

In this example it is assumed that GetTime is a function procedure which returns the value of the current time; current time is kept by an interrupt routine driven from a signal from the real-time clock. The WHILE (condition) DO loop at the beginning of the main control loop acts as a delay loop in which the program waits until it is time for the next control sample. Note that as part of the loop the variable 'time' is continually updated with the current time. At the end of the control calculation the time for the next sample is updated by adding the sample interval to the variable 'time'.

4.5 ALTERNATIVE FORMS OF THE PID ALGORITHM

In practice the simple form of the PID algorithm given in the previous section is not satisfactory and modifications have to be made to overcome its limitations. These modifications, discussed below, can be summarized under the headings:

- bumpless transfer;
- saturation (integral wind-up);
- noise; and
- modified integral and derivative calculation.

4.5.1 Bumpless transfer

Equation 4.1 implies that in the steady state, with zero error, the controlled variable M is zero. In many applications this is not true: e.g., the steady state operating conditions for a steam boiler may require the fuel line valves to be half open. In the case of the hot-air blower a non-zero voltage has to be applied to the heater input for the heater to provide heat output. It is normally the practice to modify equation

Alternative Forms of the PID Algorithm

4.1 by the addition of a constant term (MV) representing the value of the manipulated variable in the steady state, thus giving

$$m_n = K_p^* e_n + K_i^* s_n + K_d(e_n - e_{n-1}) + MV \qquad (4.7)$$

The quantity MV can be thought of as setting the operating point for the controller. If it is omitted and integral action is present, the integral action term will compensate for its omission but there will be difficulties in changing smoothly, without disturbance to the plant, from manual to automatic control. There will also be the danger that on changeover a large change (in, e.g., a valve position) will be demanded. Plant operating requirements usually demand that manual/automatic changeover be made in the so-called 'bumpless' manner. Bumpless transfer can be achieved by several means.

Method 1

The value of MV is calculated for a given steady state operating point and is inserted either as a constant in the program or by the operator just prior to the changeover from manual to automatic mode. The transfer to automatic mode is made when the value of the error is zero; at the time of changeover the integral term is set to zero and the output m equal to MV. The problem with this technique is obvious: the predetermined value of MV is only correct for one specified load; if the load is varying it may not be possible or convenient to make the changeover at the predetermined load value. If the error is not zero on changeover there will be a sudden change in the value of the manipulated variable due to the proportional action.

Method 2

In this method the value of the manipulated variable is tracked by the computer while the system is running under manual control and at the point of changeover the value of m is used to set MV, or if a predetermined value of MV is to be used, the integral sum is set as follows:

$$s = m_c - K e_c - MV$$

where m_c = value of m at changeover and e_c = the value of the error at changeover.

Method 3

This is a widely used method which is based on the use of the so-called velocity algorithm, which gives the change in the value of the manipulated variable for each sample rather than the absolute value of the variable. In continuous terms it can be obtained by differentiating, with respect to time, equation 4.1 to give

$$\frac{dm(t)}{dt} = K_c \frac{de(t)}{dt} + K_i e(t) + K_d \frac{d^2 e(t)}{dt^2} \qquad (4.8)$$

The difference equation can be obtained either by applying backward differences to equation 4.8 or by finding $m_n - m_{n-1}$ using equation 4.5 which gives

$$\Delta n = m_n - m_{n-1} = k_c \left[(e_n - e_{n-1}) + \frac{\Delta t}{T_i} e_n + \frac{T_d}{\Delta t}(e_n - 2e_{n-1} + e_{n-2}) \right] \qquad (4.9)$$

Rearranging equation 4.9 and letting $\Delta t = T_s$ gives

$$\Delta m_n = k_c \left[\left(1 + \frac{T_s}{T_i} + \frac{T_d}{T_s}\right) e_n - \left(1 + 2\frac{T_d}{T_s}\right) e_{n-1} + \frac{T_d}{T_s} e_{n-2} \right] \qquad (4.10)$$

writing

$$k_1 = k_c \left(1 + \frac{T_s}{T_i} + \frac{T_d}{T_s}\right)$$

$$k_2 = -\left(1 + 2\frac{T_d}{T_s}\right)$$

$$k_3 = T_d/T_s$$

equation 4.10 becomes

$$\Delta m_n = k_1 e_n + k_2 e_{n-1} + k_3 e_{n-2} \qquad (4.11)$$

which is easily programmed.

Because it outputs only a change in the controller position this algorithm automatically provides 'bumpless' transfer. It should be noted, however, that if a large standing error exists on changeover the response of the controller may be slow, particularly if the integral action time is long, i.e., with a large value of T_i.

Comparing the position algorithm (equation 4.6) and the velocity algorithm (equation 4.11) it can be seen that the latter is simpler to program and is inherently safer in that large changes in demanded actuator position are unlikely to occur. It is frequently the practice to limit the maximum value which Δm_n can take, thus ensuring that sudden large changes in, e.g., valve position or motor speed are avoided. These sudden changes can occur if, e.g., the measured signal is noisy or if the set point is changed; a method of dealing with noisy measurements is to use a fourth order difference algorithm to approximate de/dt and this is explained in the next section. The disturbance caused by set point changes can be reduced by modifying the algorithm to use the set point r and the measured output c rather than the error signal e.

The problem with a change in set point with the standard algorithm based on the use of error e is that the value of the set point appears in the derivative term and any change in value is differentiated, hence a sudden step change can cause a large disturbance. If in the velocity algorithm (equation 4.10) we let $e_n = r - c_n$ (note that

Alternative Forms of the PID Algorithm

r, the set point, is assumed to be constant), then the equation becomes

$$\Delta m_n = K_c \left[(c_{n-1} - c_n) + \frac{T_s}{T_i} (r - c_n) + \frac{T_d}{T_s} (2c_{n-1} - c_{n-2} - c_n) \right]$$

Changes in the set point are then accommodated by simply changing the value of the constant r.

The set point r appears only in the integral term; hence the controller must always include integral action. For security of operation a check must be included in the program to prevent the T_s/T_i parameter being set to zero or some very small value.

4.5.2 Saturation

In any practical application the value of the manipulated variable m is limited by physical constraints. A valve cannot be more than fully opened, or more than fully closed; a thyristor-controlled electric heater can supply only a given maximum amount of heat and cannot supply negative heat. If the value of the manipulated variable exceeds the maximum output of the control actuator effective feedback control is lost: good plant design should ensure that this only occurs in unusual conditions.

A simple example of what can happen is provided by considering a building heating system. The capacity of such a system is usually chosen to cope with an average winter: if extreme low temperature and high winds coincide, the system, even when operating at maximum capacity, will not be able to maintain the desired temperature. Under these conditions a large standing error in temperature will exist. If a PI controller is used, then because there is a standing error, the integral term will continue to grow; i.e., the value of s_n in equation 4.6 will be increased at each sample time. Consequently the value of the manipulated variable will increase and the demanded heat output will continually increase: since this will already be a maximum, the demand cannot be met. The changes are shown in Figure 4.3. If the wind drops and the outside temperature increases, then the building temperature will increase and eventually reach the desired temperature. The value of s_n, the integral term, will, however, still be large, since it will not be reduced until the building temperature exceeds the demanded temperature. As a consequence the integral term will continue to keep the demanded heat output at its maximum value even though the building temperature is now higher than desired.

The effect is called integral wind-up or integral saturation and results in the controller having a poor response when it comes out of a constrained condition. Many techniques have been developed for dealing with the problems of integral wind-up and the main ones are:

- fixed limits on integral term;
- stop summation on saturation;
- integral subtraction;

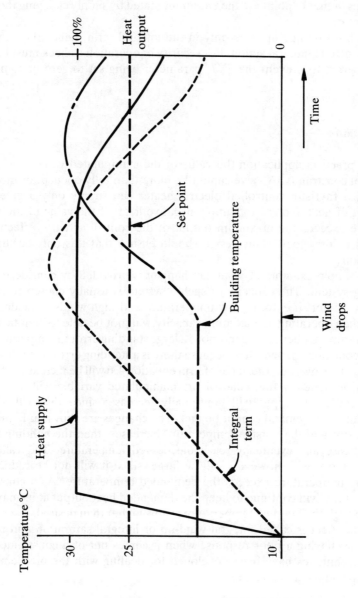

Fig. 4.3 Illustration of integral wind-up action.

Alternative Forms of the PID Algorithm

- use of velocity algorithm; and
- analytical method.

Fixed limits

A maximum and minimum value for the integral summation is fixed and if the term exceeds this value it is reset to the maximum or minimum as appropriate. The value often chosen is the maximum/minimum value of the manipulated variable; thus if $s_{max} = m_{max}$ and $s_{min} = m_{min}$ then the coding in the REPEAT ... UNTIL loop in Example 4.1 could be modified as in Example 4.6.

* * *

Example 4.6

```
REPEAT
    en := ADC; (* ADC returns value of error en *)
    s := s+en; (* integral summation *)
    IF s>smax THEN s:=smax
    ELSE IF s<smin THEN s:=smin;
    mn := kp*en + ki*s + kd*(en - enold);
    DAC(mn);
  enold := en;
UNTIL stop;
```

* * *

Stop summation

In this method the value of the integrator sum is 'frozen' when the control actuator saturates and the integrator value remains constant while the actuator is in saturation. The scheme can be implemented either by freezing the summation term when the manipulated variable falls outside the range m_{min} to m_{max} or by the use of a digital input signal from the actuator which indicates that it is at a limit.

Both of the above methods stop the integral term building up to large values during saturation but both have the disadvantage that the value of the integral term, when the system emerges from saturation, does not relate to dynamics of the plant under full power. Consequently the controller offset (provided by the integral term) lags behind the offsets required by the plant and load as the set point is reached.

Integral subtraction

The idea behind this method is that the integral value is decreased by an amount proportional to the difference between the calculated value of the manipulated variable and the maximum value allowable. The integral summation expression

$$s_n = s_{n-1} + e_n$$

is replaced by

$$s_n = s_{n-1} - K(m_n - m_{\max}) + e_n$$

The integral sum is thus decreased by the excess actuation and increased by the error. The rate of decrease is dependent on the choice of the parameter K; if it is not properly chosen then a continual saturation/desaturation oscillation can occur.

The method can be modified to stop the addition of the error part during saturation if a logic signal from the actuator indicating saturation or no saturation is available. In this case the value of the integral sum begins to decrease as soon as the actuator enters saturation and continues to decrease until it comes out of saturation, at which point integral summation begins again. The benefit of this method is that the system comes out of saturation as quickly as possible; there is, however, no attempt to match the integral term to the requirements of the plant and the value of K must be chosen by experience rather than by reference to the plant characteristics.

Velocity algorithm

It is often stated that integral wind-up can be avoided by the use of the velocity algorithm since the integral action is obtained by a summation of the increments in the output device, either at the actuator or at a device connected to the actuator, and it is this device which is subject to limiting. There is therefore an automatic integral limit which prevents a build-up of error; however, as soon as the error changes sign the actuator will come off its limit and hence at desaturation the integral term is lost.

When controllers are cascaded it must not be assumed that the use of the velocity algorithm will prevent integral wind-up. In the system shown in Figure 4.4 controllers A and B are assumed to use the velocity algorithm and both are assumed to employ PI control. Controller A is used to adjust the steam flow to a heat exchanger in order to maintain a particular water temperature in the hot water supply used to heat a room. The demanded water temperature is set by controller B. Suppose on a cold day controller B demands a water temperature of 60°C but the best the heat exchanger can do is to provide water at 55°C, the effect will be that the room temperature will remain too low and the integral sum will begin to grow and hence the set point of controller will continually be increased until it reaches its maximum limit. If the steam valve is fully open there is no action controller A can take. If now the external temperature increases there will be a delay, during which the room temperature might rise well above the desired temperature, before the set point of controller A is reduced to correct the overshoot. In order to avoid this the master controller must know when the subsidiary loop is at a limit.

Analytical approach

This method has been developed by Thomas [1984] and it uses knowledge of the plant to set the integral sum term approximately to the correct value at the point of desaturation such that the normal linear response from steady state is achieved. This is shown in Figure 4.5. For a system of the form shown in Figure 4.6, i.e., a first order

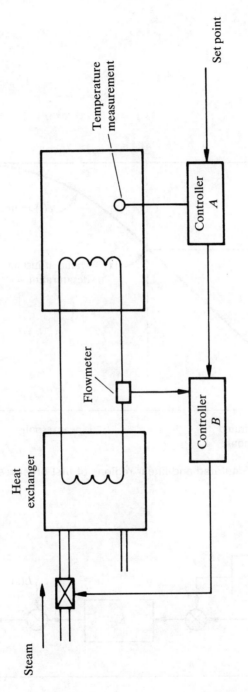

Fig. 4.4 Cascade control system.

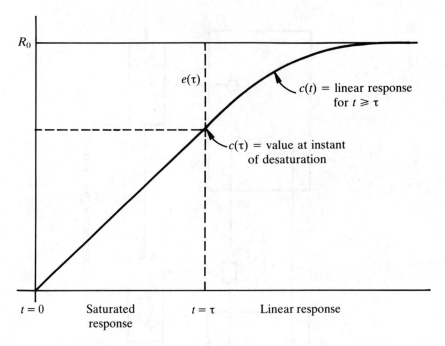

Fig. 4.5 Saturated and linear regions of first order responses.

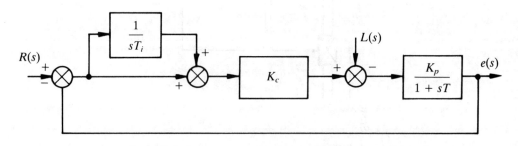

Fig. 4.6 PI control of first order plant.

Alternative Forms of the PID Algorithm

plant with a PI controller, the integral sum at the time τ when the system desaturates is given by

$$I(\tau) = \frac{1}{K_p K_c} c(\tau) + sK_p L(s) \tag{4.13}$$

If it is assumed that the load $L(s)$ is constant, or slowly varying then

$$L(s) = \frac{L_o}{s}$$

and hence equation 4.13 becomes

$$I(\tau) = \frac{1}{K_p K_c} c(\tau) + K_p L_o \tag{4.14}$$

When the control actuator desaturates the integral value should be set to the value which it would have been holding in the steady state at $c(\tau)$ and then the remaining step $e(\tau)$ will follow as in the linear case. The value of $c(\tau)$ is not known since the time, τ, of desaturation is not known; however, if prior to the control calculation, the integral term $I(t)$ is set using equation 4.14 above with the $c(t)$ for $t < \tau$ then at the instant of desaturation $I(t) = I(\tau)$. If the actuator is not in saturation then the normal integration of error takes place. This scheme is shown in the program segment below where the REAL variables 'load' and KPKC are used for $K_p L_o$ and $K_p K_c$ respectively.

* * *

Example 4.7

```
REPEAT
    cn := ADC; (* ADC returns value of error en *)
    en := cn - setpoint;
    IF mn > mnmax THEN
        s := (cn + load)/KPKC
    ELSE
     s := s+en; (* integral summation *)
    mn := kp*en + ki*s + kd*(en - enold);
    DAC(mn);
  enold := en;
UNTIL stop;
```

* * *

4.5.3 Noise

In an analog control system, a small amount of high frequency noise on the measured signal usually does not cause any problems since the dynamic components in the system act as low pass filters and attenuate the noise. If sampling is involved, high frequency noise may produce a low frequency disturbance due to folding or aliasing [see, e.g., Leigh 1985, Kuo 1980]. The low frequency disturbance has the same amplitude as the original noise and its frequency is the difference between the original noise frequency and a multiple of the sampling interval. To reduce this effect the measurement signal must be filtered before it is sampled.

For many industrial applications a simple first order filter

$$G_f(s) = \frac{1}{1 + T_f s}$$

is satisfactory. The choice of $T_f = T_s/2$ (where T_s is the sampling time for the controller) will reduce the aliasing effect by about 90% for white noise. Where T_f is small, analog filters can easily be constructed from passive elements; if T_f is greater than a few seconds combined digital and analog filtering is used. The arrangement is illustrated in Figure 4.7 and the analog filter is used to remove the high frequency noise and hence reduce the required sampling rate.

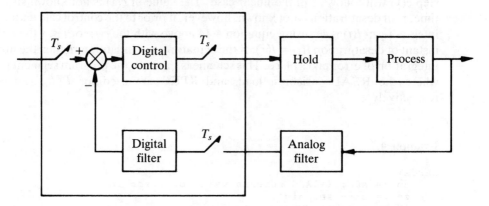

Fig. 4.7 General control system with filtering.

The numerical approximation for a first order lag

$$T_f \frac{dx}{dt} + x = u$$

with input u, output x and filter-sampling interval T_{fs} is

$$x_{n+1} = (1 - e^{-T_{fs}/T_f})x_n + e^{-T_{fs}/T_f}u_n$$

introducing $\alpha = e^{-T_{fs}/T_f}$ and applying a backwards time shift gives

$$x_n = (1 - \alpha)x_{n-1} + \alpha u_{n-1}$$

and it can be seen that if $\alpha = 1$ there is no smoothing and if $\alpha = 0$ the current measurement (input) is not used.

The sample interval T_{fs} for the input to the digital filter has to be made smaller than the time constant T_f of the filter, hence several samples of the measurement

Alternative Forms of the PID Algorithm

signal are taken for each output of the controller. The time constants of analog pre-filters are usually small and do not significantly degrade the overall performance. Excessively large values of T_f should be avoided.

As an alternative to including a separate filter algorithm the non-ideal or real PID controller (equation 4.2) which incorporates a lead/lag component instead of a derivative component is frequently used when dealing with noisy measurement signals.

4.5.4 Improved forms of algorithm for integral and derivative calculation

The positional and velocity algorithms considered so far use first and second order differences respectively to compute the derivative terms. Since differentiation or its numerical equivalent is a 'roughening' process – it accentuates noise and data errors – some form of smoothing, i.e., filtering, is required. Smoothing can be obtained by using a difference technique which averages the value over several samples; one such technique which has been used is the four-point central difference method [Bibbero 1977, Takahashi 1970]. This gives

$$\frac{de}{dt} = \frac{\Delta e}{T_s} = \frac{1}{6T_s}(e_n + 3e_{n-1} - 3e_{n-2} - e_{n-3}) \quad (4.15)$$

Substituting for $T_d/T_s(e_n - e_{n-1})$ in equation 4.5 gives

$$m_n = K_c \frac{T_d}{6T_s}(e_n + 3e_{n-1} - 3e_{n-2} - e_{n-3}) + e_n + \frac{T_s}{T_i}\sum_{k=0}^{n} e_k \quad (4.16)$$

The position algorithm using the above technique for the derivative term is thus

$$s_k = s_{k-1} + e_k$$

$$m_k = p_1 e_k + p_2 e_{k-1} - p_2 e_{k-2} - p_3 e_{k-3} + p_4 s_k$$

where

$$p_1 = 1 + \frac{K_c T_d}{6T_s}, \quad p_2 = \frac{K_c T_d}{2T_s}, \quad p_3 = \frac{K_c T_d}{6T_s}, \quad p_4 = \frac{K_c T_s}{T_i}$$

Improvements can also be made to the accuracy of the integration calculation by using the trapezoidal rule instead of the rectangular rule. If this is done, equation 4.5 can be written as

$$m_n = K_c'\left[e_n + \frac{T_s}{T_i'}\sum_{k=0}^{n}\frac{e_k + e_{k-1}}{2} + \frac{T_d'}{T_s}(e_n - e_{n-1})\right] \quad (4.17)$$

and in velocity algorithm form

$$\Delta m_n = K_c'\left[\left(1 + \frac{T_s}{2T_i'} + \frac{T_d'}{T_s}\right)e_n + \left(\frac{T_s}{2T_i'} - \frac{2T_d'}{T_s} - 1\right)e_{n-1} + \frac{T_d'}{T_s}e_{n-2}\right] \quad (4.18)$$

If this is compared with the velocity algorithm using rectangular integration we find that

$$K_c = K'_c \left(1 - \frac{T_s}{2T'_i}\right)$$

$$T_i = T'_i - T_s/2$$

$$T_d = \frac{2T'_d T'_i}{2T'_i - T_s}$$

Hence if the appropriate values for the coefficients are used the form of the two algorithms is identical.

4.6 TUNING AND CHOICE OF SAMPLE INTERVAL

Extensive work has been done on methods of tuning analog three-term controllers, beginning with the work of Ziegler and Nichols in 1942. The methods are summarized by Smith [1972 p. 169–79] and by Leigh [1985]. All of the methods are based on simple plant measurements (shown in Figure 4.8) either:

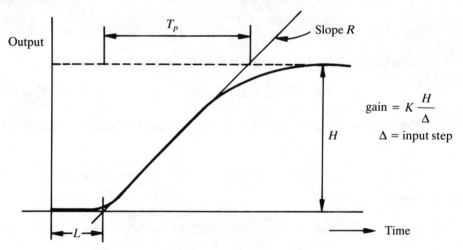

Fig. 4.8 Process reaction curve.

1. the dead time L and the slope R; or
2. dead time and the major process time constant T_p.

The assumption is that the plant can be modelled by the transfer function

$$G(s) = \frac{ke^{-Ls}}{(1 + sT_p)}$$

For the analog system the tuning problem is, given R, L or T_p, L choose K_i, K_p, K_d so as to minimize (or maximize) some performance criteria (e.g., IAE).

The digital implementation of the three-term controller involves additional variables: q the unit of quantization and T_s the sampling time; hence the performance function becomes

$$J = f(q, T_s, K_p, K_i, K_d, L, T_p)$$

The size of the quantization is not normally a problem in industrial process control since usually the control computation can be done using either real numbers or fixed point arithmetic with an adequate precision. It becomes greater in, e.g., aircraft controls and weapons systems where time constants are shorter and, in order to obtain the necessary computational speed, limited word length arithmetic has to be used. The problems are discussed extensively by Katz [1981] and Leigh [1985]; Knowles has devised a design method which takes into account quantization and sample rates [Knowles in Bennett and Linkens 1984].

Various rules of thumb have been given either for choosing a sample interval while maintaining the analog controller parameters or for modifying the controller parameters following the choice of sample interval. For example, Goff suggested that T_s, the sample interval, should be approximately $0.3L$ and that the range of $K_i = T_i/T_s$ should be:

predominantly dead time process   2
little dead time   6

and that $5 < T_d/T_s < 10$ to obtain the benefit of derivative action.

Smith [1972] has suggested that when using tuning tables based on L/T_p values the ratio $(L + T_s/2)/T_p$ should be used rather than L/T_p, in order to take into account the delay caused by sampling. He also notes that as the value of $T_s/2$ approaches the dead time, L, the performance deteriorates.

If the PID algorithm is expressed in the form given in equation 4.12 but with

$$K_p = K_c, \; K_i = K_c T_s/T_i, \; K_d = K_c T_d/T_s$$

then Takahashi [1970] gives the following rule for tuning the parameters:

$$K_p = \frac{1.2}{R(L + T_s)} - \frac{1}{2K_i}, \; K_i = \frac{0.6 T_s}{R(L + T_s/2)^2}, \; K_d = \frac{0.5}{RT_s} \text{ to } \frac{0.6}{RT_s}$$

When $T_s = 0$ the above rule converges to the standard Ziegler-Nichols result:

$$K_c = 1.2/RL, \; 1/T_i = 0.5/L, \; T_d = 0.5L$$

It should be noted that the quality of the control deteriorates when T_s increases relative to the dead time, L, and that the tuning rule fails when L/T_s is very small.

In choosing a sample interval, intuition suggests that if we increase the sampling rate such that T_s tends to zero the system will asymptotically converge towards the performance of the equivalent analog system. This is not the case, however, since the digital system has a finite resolution and hence as the sample interval decreases the change between successive values becomes less than the resolution of the system and hence information is lost. A detailed analysis of the interaction between sampling rate and word length can be found in Katz [1981].

4.7 IMPLEMENTATION OF CONTROLLER DESIGNS BASED ON PLANT MODELS

The use of detailed plant models allows a wide variety of methods to be used in the design of the controller [see, e.g., Astrom 1984, Franklin 1980, Leigh 1985, Katz 1981 and Kuo 1980]. Use of such methods gives rise to two types of representation for the controller:

- state-space representation of the difference equations; and
- transfer function in z^{-1}.

If the controller is in difference-equation form it may be programmed directly; if it is given as a transfer function it has to be realized, i.e., converted either into an electronic (or other hardware) circuit or into a computer algorithm. There are four techniques for realization:

- direct method 1;
- direct method 2;
- cascade; and
- parallel.

In terms of computer algorithms it can be shown that for a given quantization limit (i.e. word length) the cascade and parallel methods give algorithms in which the numerical errors are smaller than the errors in the algorithms produced by the two direct methods, for details the reader is referred to Katz [1980].

4.7.1 Direct methods

The transfer function can be expressed as the ratio of two polynomials in z^{-1}

$$\frac{M(z)}{E(z)} = D(z) = \frac{\sum_{j=0}^{n} a_j z^{-j}}{1 + \sum_{j=1}^{n} b_j z^{-j}} \qquad (4.19)$$

Direct method 1

The transfer function in equation 4.19 is converted directly into the difference equation

$$m_i = \sum_{j=0}^{n} a_j e_{i-j} - \sum_{j=1}^{n} b_j m_{i-j} \qquad (4.20)$$

* * *

Example

Consider a system with the transfer function

$$\frac{M(z)}{E(z)} = D(z) = \frac{3 + 3.6z^{-1} + 0.6z^{-2}}{1 + 0.1z^{-1} - 0.2z^{-2}} \qquad (4.21)$$

Implementing Controller Designs

Then by Direct method 1 the computer algorithm is simply

$$m_i = 3e_i + 3.6e_{i-1} + 0.6e_{i-2} - 0.1m_{i-1} + 0.2m_{i-2}$$

* * *

Direct method 2

In this method the difference equation is formulated by introducing an auxiliary variable $P(z)$ such that

$$\frac{M(z)}{P(z)} = \sum_{j=0}^{n} a_j z^{-j} \qquad (4.22)$$

and

$$\frac{P(z)}{E(z)} = \frac{1}{1 + \sum_{j=1}^{n} b_j z^{-j}} \qquad (4.23)$$

From equations 4.22 and 4.23 two difference equations are obtained

$$m_i = \sum_{j=0}^{n} a_j p_{i-j} \qquad (4.24)$$

and

$$p_i = e_i - \sum_{j=0}^{n} b_j p_{i-j} \qquad (4.25)$$

Using the example above the following algorithm is obtained

$$p_i = e_i - 0.1p_{i-1} + 0.2p_{i-2}$$

$$m_i = 3p_i + 3.6p_{i-1} + 0.6p_{i-2}$$

Cascade realization

If the transfer function is expressed as the product of simple block elements of first and second order as shown in Figure 4.9, then each element can be converted to a difference equation using Direct method 1 and the overall algorithm is the set of difference equations. Equation 4.21 when expressed in this form becomes

$$\frac{M(z)}{E(z)} = D(z) = \frac{3(1 + z^{-1})(1 - 0.2z^{-1})}{(1 + 0.5z^{-1})(1 - 0.4z^{-1})}$$

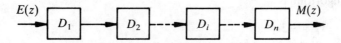

Fig. 4.9 Cascade realization.

Hence

$$D1 = 3$$
$$D2 = (1 + z^{-1})$$
$$D3 = (1 - 0.2z^{-1})$$
$$D4 = 1/(1 + 0.5z^{-1})$$
$$D5 = 1/(1 - 0.4z^{-1})$$

and letting x^1, x^2, x^3, x^4, x^5 be the outputs of blocks $D1$, $D2$, $D3$, $D4$ and $D5$ respectively the

$$x_i^1 = 4e_i$$
$$x_i^2 = x_i^1 - 0.3x_{i-1}^2$$
$$x_i^3 = x_i^2 + x_{i-1}^2$$
$$x_i^4 = x_i^3 + x_{i-1}^4$$
$$x_i^5 = x_i^4 + x_{i-1}^4$$

Parallel realization

If the transfer function is expressed in fractional form or is expanded into partial fractions then it can be represented as shown in Figure 4.10. In this case each of the transfer functions $D1$, $D2$, and $D3$ is expressed in difference equation form using Direct method 1 and the output is obtained by summing the outputs from each block.

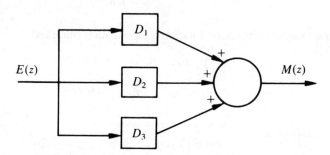

Fig. 4.10 Parallel realization.

The partial fraction expansion of equation 4.21 is

$$\frac{M(z)}{E(z)} = D(z) = -3 - \frac{1}{1 + 0.5z^{-1}} + \frac{7}{1 - 0.4z^{-1}}$$

Hence $D1 = -3$, $D2 = -1/(1 + 0.5z^{-1})$, $D3 = 7/(1 - 0.4z^{-1})$ and the algorithm is

$$x_i^1 = -3e_i$$
$$x_i^2 = -e_i - 0.5x_{i-1}^2$$
$$x_i^3 = 7e_i + x_{i-1}^3$$
$$m_i = x_i^1 + x_i^2 + x_i^3$$

4.8 THE PID CONTROLLER IN Z-TRANSFORM FORM

The PID controller can be expressed as a transfer function in z. Consider equation 4.5 and let $d = T_d/T_s$ and $g = T_s/T_i$ then

$$m_n = K_c \left[e_n + g \sum_{k=0}^{n} e_k + d(e_n - e_{n-1}) \right] \quad (4.26)$$

Since $D(z) = M(z)/E(z)$, $D(z)$ can be found by taking the z-transform of the right-hand side of equation 4.26 term by term to give

$$D(z) = K_c \left(1 + \frac{gz}{z - 1} + d - dz^{-1} \right) \quad (4.27)$$

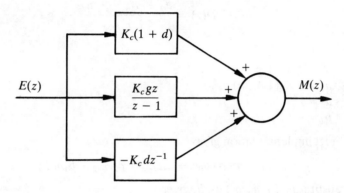

Fig. 4.11 Z-transform function form of PID controller.

Equation 4.27 represents the parallel realization of the PID controller and is shown in block form in Figure 4.11. Rewriting $K_c g/(z-1)$ as $K_c g/(1-z^{-1})$ then the algorithm is

$$x_i^1 = K_c(1 + d)e_i$$
$$x_i^2 = K_c g e_i + x_{i-1}^2$$
$$x_i^3 = -K_c d e_{i-1}$$
$$m_i = x_i^1 + x_i^2 + x_i^3$$

Substituting for d and g gives

$$x_i^1 = K_c(1 + T_d/T_s)e_i \quad (4.28a)$$

$$x_i^2 = K_c \frac{T_s}{T_i} e_i + x_i^2 \quad (4.28b)$$

$$x_i^3 = -K_c \frac{T_d}{T_s} e_{i-1}$$

and it can be seen that equation 4.28b is the integral summation term which in equation 4.6 was expressed in the form $s_n = s_n + e_n$. The algorithm from equation 4.28 is

$$s_n = K_1 e_n + s_{n-1}$$
$$m_n = K_2 e_i + K_3 e_{i-1} + s_n$$

where $K_1 = K_c T_s/T_i$, $K_2 = K_c(1 + T_d/T_s)$ and $K_3 = -K_c T_d/T_s$.

Alternatively equation 4.27 can be re-arranged to give

$$D(z) = K_c \left[\frac{(1 + g + d)z^2 - (1 + 2d)z - d}{z(z - 1)} \right]$$

Dividing numerator and denominator by z^2 gives

$$D(z) = \frac{a_0 + a_1 z^{-1} + a_2 z^{-2}}{1 + b_1 z^{-1}}$$

where

$a_0 = K_c(1 + g + d)$
$a_1 = -K_c(1 + 2d)$
$a_2 = K_c d$
$b_1 = -1$

and direct implementation gives

$$m_i = a_0 e_i + a_1 e_{i-1} + a_2 e_{i-2} - b_1 m_{i-1}$$

and substituting for a_0, a_1, a_2, b_1 gives

$$m_i = K_c \left(1 + \frac{T_s}{T_i} + \frac{T_d}{T_s}\right) e_i - K_c \left(1 + 2\frac{T_d}{T_s}\right) e_{i-1} + K_c \frac{T_d}{T_s} e_{i-2} + m_{i-1}$$

which can easily be re-arranged to give the velocity algorithm of equation 4.10.

4.9 SUMMARY

The emphasis in this chapter has been on implementing a given digital controller in a digital computer. If the controller is given in discrete form then conversion to an algorithm in a suitable form for programming is a simple task. The only difficulties arise when taking into account the practical limitations of the plant actuators and measuring instruments. If the controller is given in analog (continuous) form then considerably more thought is required, since the continuous form must be discretized.

The problem of discretization is interesting and considerably more complex than might at first be thought. There are a number of methods which can be used and these are summarized below; however, none of them exactly preserve the characteristics of the continuous system (time response, frequency response,

pole-zero locations). The main methods are:
1. impulse invariant transform (z transform);
2. impulse invariant transform with-hold;
3. mapping of differentials;
4. bilinear (or Tustin) transform;
5. bilinear transform with frequency prewarping; and
6. mapping of poles and zeroes (matched z transform).

It is not possible to give any firm indication of a best method for all applications, however, in general the bilinear transform (4 or 5) and pole-zero matching (6) give the closest approximations to the continuous system. Extensive comparisons of the methods can be found in Astrom [1984], Franklin [1980], Leigh [1985] and Katz [1980].

EXERCISES

4.1 Many personal computers have interval timers, i.e., they have a counter which can be initialized and which is incremented (or decremented) at fixed intervals by an interrupt signal. Using the technique shown in Example 4.5 write a program to output the 'bell' character (07H) at a fixed interval (e.g., 2 seconds). If you have access to a personal computer check the accuracy of the timing by using a stop watch to time a number of rings.

4.2 A person's reaction time can be measured by sending, at random intervals, a character to a VDU screen and asking the subject to press a key when the character appears. If you have access to a personal computer or some other small computer write a program to carry out such an experiment.

4.3 Modify the code of Example 4.1 to incorporate the velocity subtraction method of preventing integral action wind-up. Assume that a logic signal is available to indicate when the control actuator is in saturation.

4.4 (a) Draw a flowchart to show how bumpless transfer (Method 2 – tracking of the manipulated variable) can be incorporated into the standard PID controller.
(b) Based on the flowchart write a program (in any language) for the system.

4.5 Write a program, in any language, to implement the velocity algorithm for the PID controller.

4.6 How would you incorporate (a) into the standard PID digital controller and (b) into the velocity form of the PID controller, the requirement that the manipulated variable should not change by more than 1% between two sample intervals?

4.7 Discuss the problems of testing the computer implementation of a digital control algorithm. Work out a test scheme which would minimize the amount of time required for test purposes on the actual plant. The scheme should show the various stages of testing and should be designed to eliminate coding errors and logic design errors prior to the connection of the controller to the plant.

4.8 A digital controller has the transfer function

Using

(a) Direct method 1;
(b) Direct method 2;
(c) cascade realization; and
(d) parallel realization

find the corresponding computer algorithm.

4.9 The results of an open loop response to a unit step input for a plant are:

Time (seconds)	Output
0.1	0.01
0.2	0.02
0.3	0.06
0.4	0.14
0.5	0.24
0.6	0.34
0.7	0.44
0.8	0.54
0.9	0.64
1.0	0.71
1.1	0.76
1.2	0.79
1.3	0.80

Find (a) the approximate plant model, (b) a suitable sampling interval for a digital PID controller and (c) estimates of the optimum controller settings for PI and PID control.

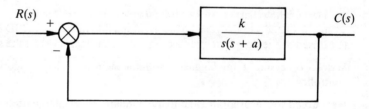

Fig. 4.12

4.10 The analog system shown in Figure 4.12 can be discretized using the z-transform plus zero-order hold method. The resulting algorithm is

$$e_n = r - c_n$$
$$c_n = (k/a^2)(Ae_{n-1} + Be_{n-2}) + (Cc_{n-1} - Dc_{n-2})$$

where
$$A = T_s.a - 1 + \exp(-a.T_s)$$
$$B = 1 - \exp(-a.T_s) - T_s.a.\exp(-a.T_s)$$
$$C = 1 + \exp(-a.T_s)$$
$$D = \exp(-a.T_s)$$
$$T_s = \text{sampling interval}$$

Write a program which will enable you to calculate the change in output of the system (c_n) with time. It is suggested that 50 values are calculated. The program should enable different values of k, a, T_s and r to be entered.

4.11 Using the program of Example 4.10, set $k = 2$, $a = 1$, $r = 1$ and investigate the response of the system for different values of T_s. It is suggested that $T_s = 0.02, 0.05, 0.1, 0.2, 0.5$). Compare the results (e.g., in terms of maximum overshoot) with the exact solution for the continuous system (maximum overshoot = 30.5%).

REFERENCES AND BIBLIOGRAPHY

AHSON, S.I., et al. (1983). 'A microprocessor-based multi-loop process controller', *IEEE Trans. on Industrial Engineering*, **IE- 30(1)**, pp. 34–9

ASTROM, K.J. and WITTENMARK, B. (1984), *Computer Controlled Systems*, Prentice Hall

AUSLANDER, D.M., TAKAHASHI, Y. and TOMIZUKA, M. (1978), 'Direct digital process control: practice and algorithms for microprocessor application', *Proc. IEEE*, **66**, pp. 199–208

AUSLANDER, D.M. and SAGUES, P. (1981), *Microprocessors for Measurement and Control*, Osborne/McGraw-Hill

BENNETT, S. and LINKENS, D.A. (eds.) (1982), *Computer Control of Industrial Processes*, Peter Peregrinus, Stevenage

BENNETT, S. and LINKENS, D.A. (eds.) (1984), *Real-time Computer Control*, Peter Peregrinus, Stevenage

BIBBERO, R.J. (1977), *Microprocessors in Instruments and Control*, Wiley

CASSELL, D.A. (1983), *Microcomputers and Modern Control Engineering*, Reston Publishing Co

COHEN, G.H. and COON, G.A. (1953), 'Theoretical consideration of retarded control', *Trans. ASME*, **75**, p. 827

FRANKLIN, G.F. and POWELL, J.D. (1980), *Digital Control of Dynamic Systems*, Addison Wesley

GOFF, K.W. (1966), 'A systematic approach to DDC design', *ISA Journal*, p. 44–54

HINE, D. and BURBRIDGE, L. (1979), 'A microcomputer algorithm for open-loop step-motor control', *Trans. IMC*, **1**, pp. 233–9

ISERMAN, R. (1981), *Digital Control Systems*, Springer

JOHNSON, C.D. (1984), *Microprocessor-based Process Control*, Prentice Hall

KATZ, P. (1981), *Digital Control using Microprocessors*, Prentice Hall

KNOWLES, J.B. (1984), 'Some DDC design procedures', in Bennett (1984)

KUO, B.C. (1980), *Digital Control Systems*, Holt Saunders

LEIGH, J.R. (1985) *Applied Digital Control*, Prentice Hall

MORONEY, P., WILLSKY, A.S. and HOUPT, P.K. (1980), 'The digital implementation of control compensators: the coefficient wordlength issue', *IEEE Trans. Automatic Control*, **AC-25(4)**, pp. 621–30

MELLICHAMP, D.A. (ed.), (1983), *Real-time Computing with Applications to Data Acquisition and Control*, Van Nostrand

SAVAS, E.S. (1965), *Computer Control of Industrial Processes*, McGraw-Hill

SMITH, C.L. (1972), *Digital Computer Process Control*, Intext Educational

STEPHANOPOULOS, G. (1985), *Chemical Process Control*, Prentice Hall

TABAK, D. (1983), 'Hardware and software aspects of control applications', in Tzafestas (1983)

TAKAHASHI, Y., RABINS, M.J. and AUSLANDER, D.M. (1970), *Control and Dynamic Systems*, Addison Wesley

THOMAS, H.W., SANDOZ, D.J. and THOMSON, M. (1983), 'New desaturation strategy for digital PID controllers', *Proc. IEE*, **130(4)**, pp. 188–92

TZAFESTAS, S.G. (ed.) (1983), *Microprocessors in Signal Processing, Measurement and Control*, Reidel, Dordrecht

ZIEGLER, J.G. and NICHOLS, N.B. (1942), 'Optimum settings for automatic controllers', *Trans. ASME*, **64**, pp. 759–68

5

Design of Real-time Systems

5.1 GENERAL APPROACH

The general approach to the design of real-time computer systems is no different in outline from the approach required for any computer based system. The work can be divided into two main sections:

- planning phase; and
- development phase.

The planning phase is illustrated in Figure 5.1 and is concerned with interpreting user requirements to produce a detailed specification of the system to be developed and an outline plan of the resources – people, time, equipment, costs – required to carry out the development. At this stage preliminary decisions regarding the division of functions between hardware and software will be made. A preliminary assessment of the type of computer structure – a single central computer, a hierarchical system, or a distributed system – will need to be made. The outcome of this stage is a *specification* or *requirements* document. (The terminology used in books on software engineering can be confusing; some refer to a specification requirements document as well as to specification and requirements documents. It is clearer and simpler to consider that documents produced by the user or customer describe requirements and documents produced by the supplier or designer give the specifications.)

It cannot be emphasized too strongly that the specification documents for both the hardware and software which result from this phase must be complete, detailed and unambiguous. General experience has shown that a large proportion of 'errors' which appear in the final system can be traced back to unclear, ambiguous or faulty specification documents. There is always a strong temptation to say 'It can be decided later'; deciding it later can result in the need to change other parts of the system which have already been designed and in changing them errors are introduced.

It is frequently recommended that, as part of the specification of the system, the user documentation should also be produced. Whatever the form of the documentation produced it should have been checked in detail with the customer.

The stages of the development phase are shown in Figure 5.2. The aim of the preliminary design stage is to decompose the system into a set of specific sub-tasks which can be considered separately. The preliminary design stage is also referred to as the high-level design stage. The inputs to this stage are the high-level specifications; the outputs are the global data structures and the high-level software

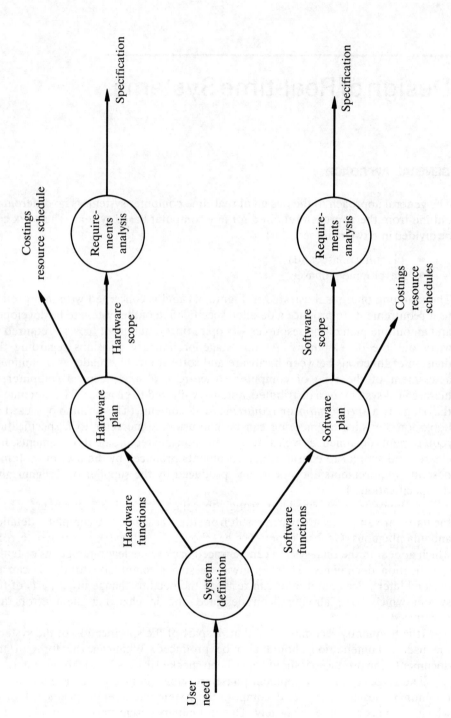

Fig. 5.1 Planning phase.

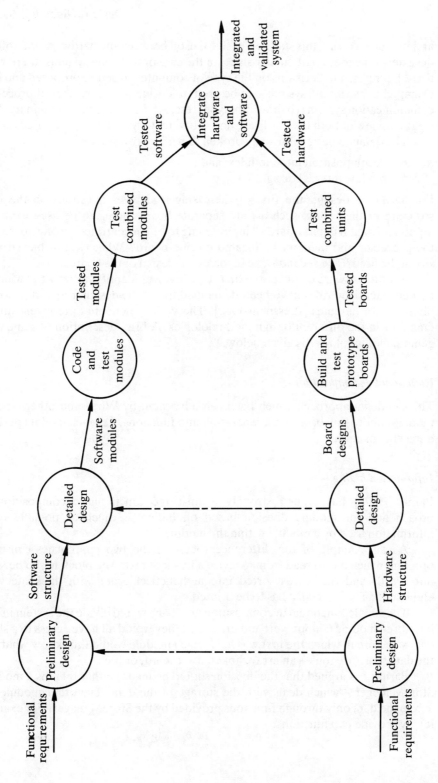

Fig. 5.2 Development phase.

architecture. During this stage extensive liaison between the hardware and software designers is needed, particularly since, in the case of real-time systems, there will be a need to revise the decisions on the type of computer structure proposed and if, for example, a distributed system is to be used, to decide on the number of processors, communication systems (bandwidth, type), etc. At the end of the preliminary design stage a review of both the hardware and software designs should be carried out.

The detailed design is usually broken down into two stages:

- decomposition into modules; and
- module internal design.

For hardware design, the first of these stages involves questions on the board structure of the system such as: are separate boards going to be used for analog inputs and digital inputs or are all inputs going to be concentrated on one board? Can the processor and memory be located on one board? What type of bus structure should be used? The second stage involves the design of the boards.

For the software engineer the first stage involves identifying activities which are related. Heuristic rules have been developed to aid the designer with decisions on division into modules [Pressman 1982]. The various heuristics are given differing emphasis in the different design methodologies. A brief description of some of the general methodologies is given below.

Functional decomposition

The top-down approach which has been advocated by Wirth and others leads to module subdivision based on a separation into functions, i.e. each module performs a specific function.

Information hiding

Parnas [1972] has argued strongly against the functional decomposition approach and for module division based on hiding as much as possible of the information used by a module within the module.

As an example of the differences between the two approaches consider a program which has to read from a device a block of text. The block has to be sorted into words and the words sorted into alphabetical order with duplicate words eliminated. The sorted list has to be printed.

The simple functional decomposition approach would divide the system initially into three modules: input, sort and print, and they would all have access to a shared data structure in which the text was held. Each module would thus know what form the data structure took – an array, linked list, record, or file.

Parnas has argued that the division should be made so that there is a module – StoreManager – which deals with the storage of the data. The other modules can access the data only through functions provided by the StoreManager. For example, it may provide two functions:

- Put (word); and
- Get (word);

to enable information to be put into store and retrieved from store. The other modules need not know how the storage is organized. The advantage claimed for this approach is that design changes to one module do not affect another module.

Coupling and cohesion

The maximizing of module cohesion and minimizing of coupling between modules are the heuristics underlying data-flow design methodologies. The heuristic and methodologies have been developed by Constantine [1978], Myers [1978] and Stevens [1974].

Partition to minimize interfaces

This heuristic has been proposed by DeMarco [1973] and can be combined with data-flow methods. It suggests that the transformations indicated on the data-flow diagram should be grouped so as to minimize the number of interconnections between the modules.

For real-time systems additional heuristics are required, one of which is to divide modules into the following categories:

- real-time, hard constraint;
- real-time, soft constraint; and
- interactive.

The arguments given in Chapter 1 regarding the validation of different types of program would suggest a rule which aims to minimize the amount of software which falls into the hard constraint category since this is the most difficult to design and test.

In terms of software design the major differences between real-time and standard systems occur in the preliminary design and decomposition into modules stages of the design procedure and this chapter will concentrate on these areas. Module internal design, coding, and testing follows a similar pattern to that for non-real-time software.

5.2 SPECIFICATION DOCUMENT

To provide an example for the design procedures being described we shall consider the hot-air blower system described in Chapter 1. It is assumed that the planning phase has been completed and a specification document has been prepared. A shortened version of such a document is given below.

Example 5.1 Hot-air blower specification

Plant interface

Input from plant:
   Outlet temperature: analog signal, range 0–10 V, corresponding to 20°C to 64°C, linear relationship.
Output to plant:
   Heater control: analog signal 0 V to -10 V corresponding to full heat (0V) to no heat (-10 V), linear relationship.

Control

A PID controller with a sampling interval of 40 ms is to be used. The sampling interval may be changed, but will not be less than 40 ms. The controller parameters are to be expressed to the user in standard analog form i.e. proportional band, integral action time and derivative action time. The set-point is to be entered from the keyboard. The controller parameters are to be variable and are to be entered from the keyboard.

Operator communication

Display
The operator display is as shown below:
   Set temperature   : $nn.n$ °C   Date : $dd/mm/yyyy$
   Actual temperature : $nn.n$ °C   Time : $hh:mm$
   Error   : $nn.n$ °C
   Heater output   : nn %FS   Sampling interval : nn ms

Controller settings
   Proportional band : nnn %
   Integral action   : $nn.nn$ s
   Derivative action : $nn.nn$ s
The values on the display will be updated every 5 s.

Operator input
The operator can at any time enter a new set point or new values for the control parameters. This is done by pressing the 'ESC' key. In response to 'ESC' a menu is shown on the bottom of the display screen.

   1. Set temperature = $nn.n$ 2. Proportional band = nnn %
   3. Integral action = $nn.nn$ 4. Derivative action = $nn.nn$
   5. Sampling interval = nn 6. Management information
   7. Accept entries
   Select number of item to change ⟩

Preliminary Design

In response to the number entered, the present value of the item selected will be deleted from the display and the cursor positioned ready for input of new value. The process will be repeated until Item 7 – Accept entries is selected at which time the bottom part of the display will be cleared and the new values shown in the top part of the display.

Management information

On selection of Item 6 of the operator menu a management summary of the performance of the plant over the previous 24 hours will be given. The summary provides the following information:

1. Average error in °C in 24 hour period.
2. Average heat demand %FS in 24 hour period.
3. For each 15 minute period:
   (a) average demanded temperature;
   (b) average error; and
   (c) average heat demand.
4. Date and time of output.

General information

There will be a requirement for a maximum of 12 control units. A single display and entry keyboard which can be switched between the units is adequate.

5.3 PRELIMINARY DESIGN

5.3.1 Hardware design

The number of units is not sufficient to warrant a specially designed processor and the most likely economic solution is one based either on a single board computer or on a modular system using a standard bus. The requirements should be easily satisfied by an 8-bit microprocessor. A single channel analog-to-digital converter and a single channel digital-to-analog converter will be required. A decision will have to be made as to the speed and resolution of conversion of each. It is also unlikely that the use of multiple processors would be economically viable. Two possible configurations which might be examined in detail and costed are shown in Figure 5.3.

Configuration 1 provides a simple hardware solution, with the minimum number of processors and a simple mechanical switch to connect the keyboard and display to each processor in turn. However, each processor will have to contain display and input programs and management information programs.

Configuration 2 allows for separation of actual control from the management and operator display/input programs, but it requires additional hardware in the form of an extra processor and a communication network; there will, however, be a saving in the amount of memory required on each of the control processors.

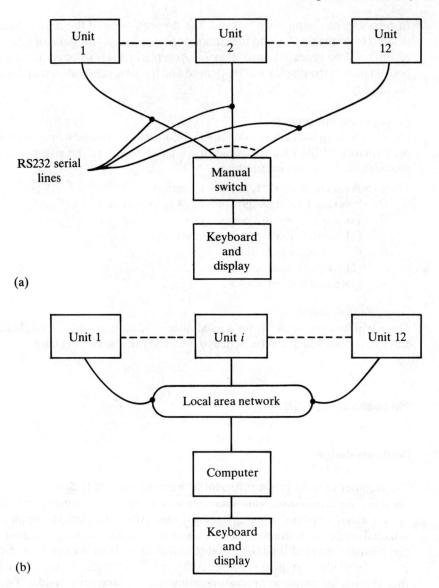

Fig. 5.3 Possible hardware configurations. (a) Configuration 1; (b) configuration 2.

Part of the hardware outline design would be to evaluate these two – and possibly other – approaches.

Since there is a requirement for date and time to be available a real-time clock will be required. The minimum sampling interval is given as 40 ms and hence a clock based on the line frequency will be adequate since at 50 Hz it will give an interrupt every 20 ms.

Preliminary Design

A decision will have to be made as to whether to use a system including a video generator or to use a standard terminal. Since only one display is required for up to 12 systems it may be cheaper to use a standard terminal connected via a serial interface rather than providing a video generator and memory in each unit. This is a decision which would depend on a careful analysis of the costs and benefits of each approach.

5.3.2 Software design

Based on the categories outlined above, the software divides into the segments shown in Figure 5.4. The control module has a hard constraint in that it must run every 40 ms. In practice this constraint may be relaxed a little to be, say, 40 ms $\pm$ 1 ms with an average value over 1 minute of say 40 ms $\pm$ 0.5 ms. In general the sampling time can be specified as $T_s \pm e_s$ with an average value over time T of $T_s \pm e_a$. The requirement may also be relaxed to allow, e.g., one sample in 100 to be missed. These constraints will form part of the test specification and will be needed during testing of the system.

The requirement on the clock/calendar module is that it must run every 20 ms in order not to miss a clock pulse. This constraint can be changed into a soft constraint if

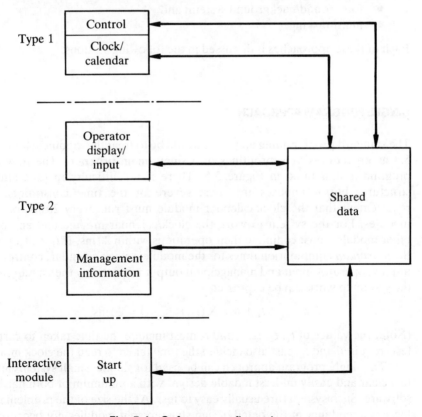

Fig. 5.4 Software configuration.

some additional hardware is provided in the form of a counter which can be read and reset by the clock/calendar module. The constraint could now be, say, an average response time of one second with a maximum interval between reading the counter of 5 seconds. (For these values what size of counter would be required?)

The operator display, as specified, has a hard constraint in that an update interval of 5 seconds is given. Commonsense would suggest that this need not be a hard constraint (an average time of 5 seconds should be adequate) but a maximum time would also have to be specified: say, 10 seconds.

Similarly soft constraints are adequate for operator input and for management information logs. These would have to be decided upon and agreed with the customer. They should form part of the specification in the requirements document.

The start-up module does not have to operate in real time and hence can be considered as a standard interactive module.

There are obviously several different activities which can be divided into sub-problems. The sub-problems will have to share a certain amount of information and how this is done and how the next stages of the design proceed will depend upon the general approach to the implementation. There are three possibilities:

- single program;
- foreground/background system; and
- multi-tasking.

Each of these approaches is discussed in the following sections.

5.4 SINGLE PROGRAM APPROACH

The standard programming approach would be to treat the modules shown in Figure 5.4 as procedures or subroutines of a single main program. The flow of such a program is illustrated in Figure 5.5. There is no difficulty in programming this structure, but it imposes the most severe of the time constraints, i.e., the requirement that the clock/calendar module must run every 20 ms, on all of the modules. For the system to work the clock/calendar module and any one of the other modules must complete their operations within 20 ms. If t_1, t_2, t_3, t_4 and t_5 are the *maximum* computation times for the module's clock/calendar, control, operator display, operator input and management output respectively then a requirement for the system to work can be expressed as

$$t_1 + \mathrm{MAX}\,(t_2, t_3, t_4, t_5) \leq 20\,\mathrm{ms}$$

(Note: the values of t_1, t_2, t_3, t_4 and t_5 must include the time taken to carry out the tests required and t_1 must also include the time taken to read the clock input line.)

The single program approach can be used for simple, small systems and it leads to a clear and easily understandable design, with a minimum of both hardware and software. Such systems are usually easy to test. As the size of the problem increases, there is a tendency at the detail design stage to split modules not because they are

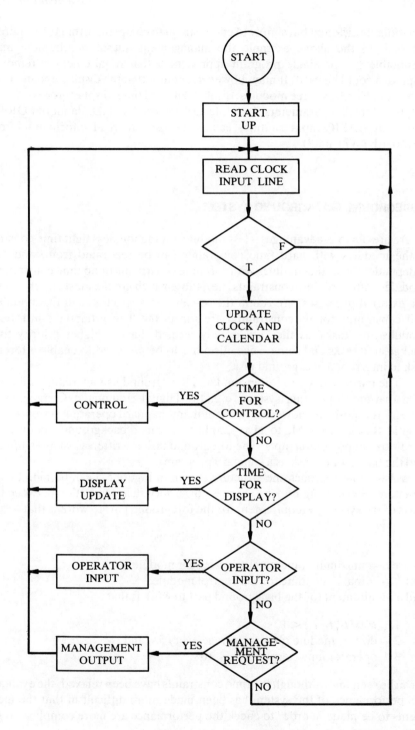

Fig. 5.5 Single program approach.

functionally different but simply to enable them to complete within the required time interval. In the above example the management output requirement makes it unsuitable for the single program approach; if that requirement is removed the approach could be used. It may, however, require that the display update module is subdivided into say three modules: display date and time; display process values; and display controller parameters. It should also be noted that the standard FORTRAN, Pascal or BASIC input routines cannot be used to read information from the keyboard. (Why not?)

5.5   FOREGROUND/BACKGROUND SYSTEM

There are obvious advantages – less module interaction, less tight time constraints – if the modules with hard time constraints can be separated from, and handled independently of, the modules with soft time constraints or no time constraints. The modules with hard time constraints, i.e., those which are the most closely coupled to the external processes, are run in the so-called 'foreground' and the modules with soft constraints (or no constraints) are run in the 'background'. The foreground modules, or 'tasks' as they are usually termed, have a higher priority than the background tasks and some mechanism has to be provided to enable a foreground task to interrupt a background task.

The partitioning into foreground and background usually requires the support of a real-time operating system, e.g. the Digital Equipment Company's RT/11 system. It is possible, however, to adapt many standard operating systems, e.g., the Digital Research CP/M, to give simple foreground/background operation if the hardware supports interrupts. The foreground task is written as an interrupt routine and the background task as a standard program.

Using this approach the structure shown in Figure 5.5 can be modified to the structure shown in Figure 5.6. There is now a very clear separation between the two parts of the system; a requirement for the foreground part to work is that

$$t_1 + t_2 \leq 20\,\text{ms}$$

where t_1 = maximum time for clock/calendar module
and t_2 = maximum time for the control module;
and a requirement for the background part to work is that

1. $\max(t_3, t_4, t_5) < 10\,\text{s}$;
2. display module runs on average every 5 s; and
3. operator input responds in $< 10\,\text{s}$.

It can be seen that, although the time constraints have been relaxed, the evaluation of the performance of the system has been made more difficult in that the measurements to be made in order to check the performance are more complicated than in the single program case.

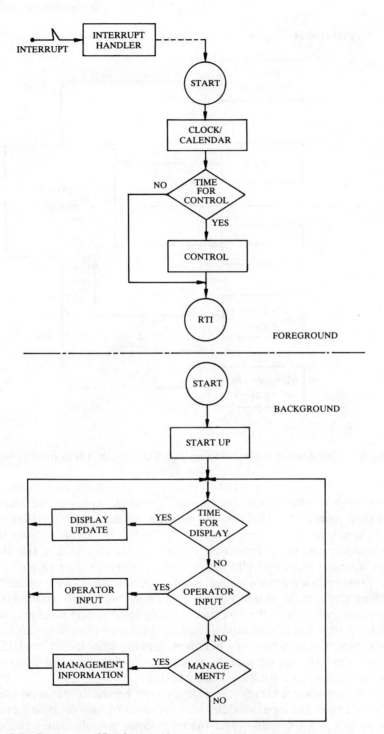

Fig. 5.6 Foreground/background approach. (a) Foreground; (b) background.

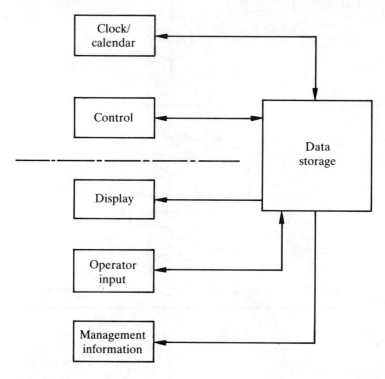

Fig. 5.7 Software modules for foreground/background system showing data storage.

Although the foreground/background approach separates the *control* structure of the foreground and background modules, the modules are still linked through the data structure as is shown in Figure 5.7. The linkage occurs because they share data variables; e.g., in the hot-air blower system, the control task, the display task and the operator input task all require access to the controller parameters. In the single program (sometimes called single-tasking) there was no difficulty in controlling access to the shared variables since only one module (task) was active at any one time, whereas in the foreground/background system tasks may be used in parallel, i.e., one foreground module and one background module may be active at the same time. (Note: active does not mean 'running' since if only one CPU is being used only one task can be using it at any instant, but both the foreground and background tasks may have the potential to run.)

In this particular example the variables can be shared between the control, display, and operator input modules without any difficulty since only one module writes to any given variable. The operator input module writes the controller parameters and set point variables, the clock/calendar module writes to the date and time variables and the control module writes to the plant data variables (error and output temperature). There is, however, one precaution which must be taken: that is, the input from the operator must be buffered and only transferred to the shared storage when it has been verified. Example 5.2 shows a method of doing this.

* * *

Example 5.2 Buffering of parameter input data

```
MODULE HotAirBlower;
VAR
   p1,p2,p3 : REAL; (* Contoller parameters declared as
                       global variables *)
PROCEDURE GetParameters(VAR x,y,z : REAL);
  BEGIN
   ...
   (* get new parameters from terminal and store in x,y,z *)
   ...
END GetParameters ;
PROCEDURE OperatorInput;
VAR
   x,y,z : REAL;
  BEGIN
   GetParameters(x,y,z);
   (* insert code to verify here *)
   p1:=x;   (* transfer parameters to global variables *)
   p2:=y;
   p3:=z;
END OperatorInput;
BEGIN
  (* main program *)
END HotAirBlower.
```

* * *

To understand the reasons for buffering, let us consider what would happen if, when a new value was entered, it was stored directly in the shared data areas. Suppose the controller was operating with P1 = 10, P2 = 5 and P3 = 6 and it was decided that the new values of the control parameters should be P1 = 20, P2 = 3 and P3 = 0.5. As soon as the new value of P1 is entered the controller begins to operate with P1 = 20, P2 = 5, P3 = 6, i.e., neither the old nor the new values. Much harm may not be done if the operator enters the values quickly, but what happens if, after the operator has entered P1, the telephone rings or he/she is interrupted in some other way and consequently forgets to complete the entry? The plant could be left running with a completely incorrect (and possibly unstable) controller.

In fact the method used in Example 5.2 is not strictly correct since an interrupt could occur between transferring x to p1 and y to p2, in which case an incorrect controller would be used. For a simple feedback controller this would have little effect since it would be corrected on the next sample. It may be more serious if the change were to a sequence of operations. The potential for serious and possibly dangerous consequences is not great in small, simple systems (a good reason for keeping systems small and simple whenever possible): it is much greater in large systems.

The transfer of data between the foreground and background tasks, i.e., the statements:

```
p1:=x;
p2:=y;
p3:=z;
```

form what is known as a *critical section* of the program and should be an indivisible action. There is a simple way of ensuring this: it is to inhibit all interrupts during the transfer, i.e.

```
InhibitInterrupts;
p1:=x;
p2:=y;
p3:=z;
EnableInterrupts;
```

The difficulty with this simple approach is that it is undesirable for several separate modules each to have access to the basic hardware of the machine and each to be able to change the status of the interrupts. It has been found by experience that modules concerned with the details of the computer hardware are difficult to design, code and test and have a higher error rate than the average. Hence it is good practice to limit the number of such modules.

Ideally transfers should take place at a time suitable for the controller module, i.e., the operator module and the controller module should be synchronized or should rendezvous. Methods of achieving synchronization are discussed in Chapter 7.

5.6 MULTI-TASKING APPROACH

The design and programming of large real-time (and large interactive) systems can be considerably eased if the foreground/background partitioning can be extended into multiple partitions to allow the concept of many active tasks. This allows, at the preliminary design stage, the idea of considering each activity that has to be carried out as a separate task. (Computer scientists use the word *process* rather than task but this usage has not been adopted because of the possible confusion which could arise between internal computer processes and the external processes on the plant.) The implications of this approach are that each task may be carried out in parallel and there is no assumption made at the preliminary design stage as to how many processors will be used in the system.

The implementation of a multi-tasking system requires the ability to

- create separate tasks;
- schedule running of the tasks, usually on a priority basis;
- share data between tasks;

Real-time Software Design: Approach

- synchronize tasks with each other and with external events;
- prevent tasks corrupting each other; and
- control the starting and stopping of tasks.

These facilities are typically provided by a real-time operating system and the approach to implementing a multi-tasking system using a real-time operating system is considered in Chapter 6.

5.7 GENERAL APPROACH TO REAL-TIME SOFTWARE DESIGN

In the previous sections the technical details of implementing the design were allowed to intrude on the design of the software architecture. This occurred because the notations being used to represent the design were not adequate and did not support the necessary abstract entities.

The design of large complex systems (it is assumed that if multi-tasking is to be used the system will be large and complex) requires the creation of two models. The abstract model – sometimes referred to as the logical or essential model – which contains no information on the implementation; and the production or implementation model which contains technical details of how the system is to operate (see Figure 5.8). The two models may be developed serially; i.e., the abstract model is developed and then the implementation model, but it is more usual to develop the

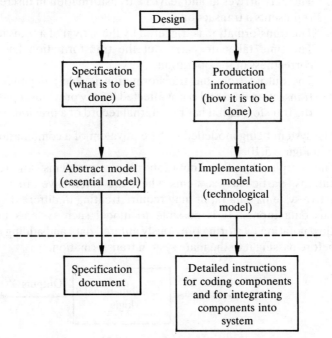

Fig. 5.8 Abstract and implementation models.

Fig. 5.9   Data flow – basic notation.

two models in parallel since high-level implementation decisions may influence lower level abstract model decisions.

The development of a specification-based abstract model is dependent on having a design notation which provides a rich enough set of conceptual ideas for the full expression of all the design requirements. It must also be accompanied by a methodology which enables the abstract model to be translated into a production or implementation model. Guidance in developing the design can be provided by design heuristics as discussed earlier.

The most suitable notations and methodologies for real-time design are the ones which have been developed from the data-flow design approach of Myers [1974]. The underlying assumption of the original data-flow technique is that a system or sub-system can be represented by a data transformation, shown on the diagram as a bubble (see Figure 5.9). The data flows are represented by arrows. Implicit in the representation are the following assumptions.

1. The data arrives at and leaves a transformation in discrete units: each unit represents a transaction.
2. The transformation is triggered by the arrival of a transaction.
3. The time taken to carry out the transformation has no effect on the correctness of the operation.
4. The output from the transformation is solely dependent on the current transaction data and is not affected by any previous input. This implies that the transformation has no internal record of a previous transaction.

The system being modelled is the equivalent of a combinational logic circuit as shown in Figure 5.10.

The simple data-flow notation causes problems for representing many applications, particularly systems which are interactive. In a large majority of interactive systems data input may require treating in different ways depending on previous data inputs. It is possible to model such systems using the standard data-flow notation by storing previously entered data and adding it to newly entered data before passing it to the main system transformation.

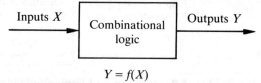

Fig. 5.10   Combinational logic equivalent of simple data flow.

Real-time Software Design: Approach

As an example of this approach consider a word-processing system in which characters entered from the keyboard may be either text or part of a command string. The start of a command string is indicated by entering the 'ESC' character and the end of the string is indicated by the 'CR' character. A possible solution is shown in Figure 5.11 in which an input transformation reads a character from the display and prefixes it with a character taken from the command character store. The command start character (ESC) and end character (CR) are stored by the input transformation in the store. The data received by the main transformation is thus either ESC + character, or CR + character, and the word-processing transformation interprets the data accordingly.

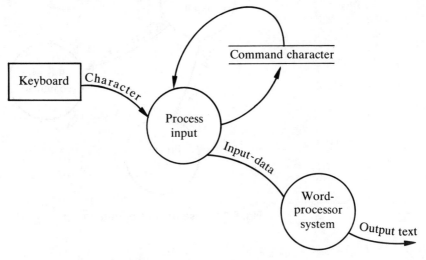

Fig. 5.11 Command processing.

An alternative approach is to model the system such that the input transformation remembers that a start command character has been received by changing its internal state such that subsequent characters received are treated as commands and passed on to a command processor rather than a text processor. This type of model is shown in Figure 5.12 and additional notation is introduced to indicate a 'control' transformation rather than a data transformation. Control actions are indicated by dotted lines rather than solid lines. The control flows can be considered as events which do not have data associated with them. The control transformation would be specified by means of a state transaction table.

In the second approach the system is model as being equivalent to a sequential logic system as illustrated in Figure 5.13. The output is now dependent not only on the current inputs, but on the internal state of the system as influenced by previous inputs as well.

If real-time systems are examined we find that there is a need to represent additional features. Typically, a real time system will have a range of different types of input and output which require notations to represent:

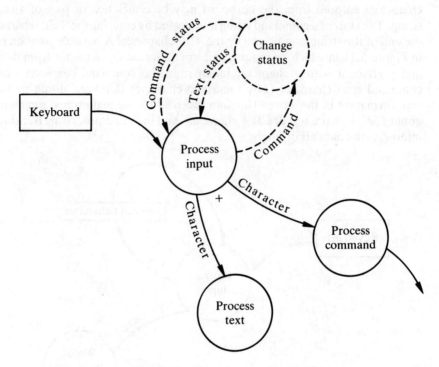

Fig. 5.12 Use of control notation.

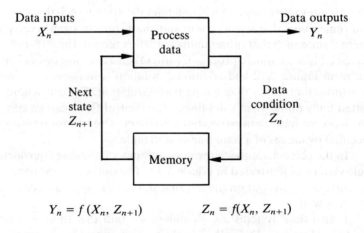

$Y_n = f(X_n, Z_{n+1})$ $\qquad$ $Z_n = f(X_n, Z_{n+1})$

Fig. 5.13 Sequential system.

1. discrete events which can occur at any time;
2. continuously changing data values; and
3. discrete data values (equivalent of transactions).

Transformations may need to be initiated:

1. on the arrival of a discrete data value;
2. at predetermined time intervals;
3. by specific internally or externally generated events; or
4. continuously.

Changes to the internal state of the system may be caused not only by data conditions but also by external events. Hatley [1984] has suggested a model as shown in Figure 5.14. The control inputs and outputs are event flows.

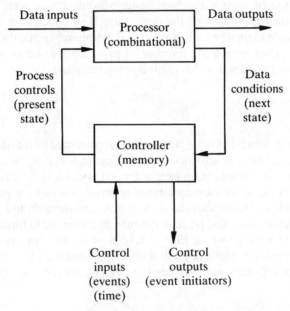

Fig. 5.14 General model of a real-time system (after Hatley 1984).

To deal with some of the above a number of real-time design methodologies have been developed. The main ones are summarized below.

1. MASCOT (modular approach to software construction operation and test); introduced by Jackson and Simpson [1975].
2. DARTS (design approach for real-time systems); Gomaa [1984].
3. PAISLey (process-oriented, applicative, interpretable, specification language) Zave [1983].
4. A structured analysis method for large, real-time systems; Hatley [1984].
5. Structured development for real-time systems; Ward and Mellor [1986].

In the sections which follow two of the methods, MASCOT and Ward and Mellor's structured development, will be considered.

5.8 MASCOT

The official definition of MASCOT is given in *The Official Handbook of MASCOT*, MASCOT Suppliers Association. Discussion of various aspects of the method can be found in Jackson and Simpson [1975], Simpson and Jackson [1978] and Simpson [1982]. Using the MASCOT system, preliminary design can be interpreted as constructing a virtual machine to solve the problem. The virtual machine is built from a particular set of abstract entities [Allworth 1981]. The construction of such a machine using Modula-2 has been described by Budgen [1982]. The later stages of the design are then concerned with implementing the virtual machine on a real, or several real, computers since the design method makes no assumptions about the number of processors to be used. The set of abstract entities used are discussed in the following sections.

5.8.1 Activity

Each activity which has to be performed is represented by a circle or bubble as shown in Figure 5.15. Thus the activities shown in Figure 5.4 could be represented as shown in Figure 5.16. The virtual machine is assumed to be able to treat each of these activities as a separate task, so they are sometimes referred to as tasks or processes. However, it is preferable to call them activities since at this stage there should not be a decision on the actual task structure: several activities may be combined to form a single task.

The system shown in Figure 5.16 is of course unworkable since the various activities perform interrelated functions and must therefore communicate with each other. The activities may also need to synchronize their actions.

Fig. 5.15 Notation for an activity.

5.8.2 Communication

Communication between activities can be divided into three types:
- direct transfer between two activities;
- shared information between several activities; and
- synchronization signals.

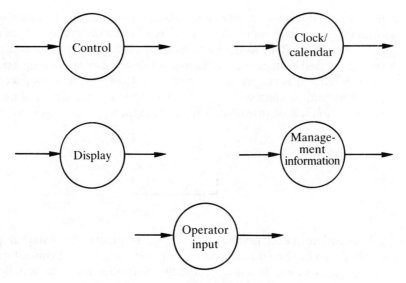

Fig. 5.16 Activities for hot-air blower.

For example, in the hot-air blower system the clock/calendar activity would send time information directly to the control activity, whereas the process values and the controller parameters would be shared between the control, display, operator input, and management output activities.

5.8.3 Channels

A useful concept for describing direct communication between activities is the *channel* (this is similar to the *pipe*). A channel, besides connecting two activities, may also have some storage associated with it so that more than one item of information can pass through the channel at any one time. Items will normally be ordered so that the first item passed into a channel will be the first item to be removed at the other end. It is unnecessary at the design stage to be concerned with how a channel might be implemented (some methods are described in Chapter 7). A channel is represented by the symbol shown in Figure 5.17.

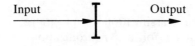

Fig. 5.17 A channel.

5.8.4 Pools

The name *pool* is given to a collection of information which is available for reading and/or writing by a number of activities in the system. The operation of reading an item in a pool does not remove or change the item; hence information stored in a pool may be read many times. A pool acts as a databank or repository of information and any item in it will be available with equal facility to all activities using the pool. Again, at the design stage the mechanism for creating and operating a pool need not be considered – it is simply an abstract idea. However, it is assumed that in any implementation, mechanisms to protect corruption of the data and ordered use of the information will be provided. The notation for a pool is shown in Figure 5.18.

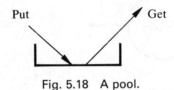

Fig. 5.18 A pool.

Physical devices (filestores, displays), including the external plant being controlled, can be thought of as pools. For example, a simple control system can be represented as shown in Figure 5.19; the controller reads information from the temperature measuring bridge, processes it and writes to the heater control. The software model treats the temperature measuring bridge and the heater control as pools and the controller as an activity.

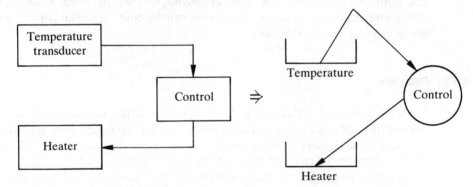

Fig. 5.19 A simple temperature control system illustrating the design notation.

5.8.5 Synchronization

In discussing the classification of programs, based on the requirements for validation, the need for a synchronization procedure in order to validate a multi-tasking program, in which the data was shared between tasks, was stressed. Activities also need to be able to *stop*, *start* and *delay* themselves and other activities. They also need to be able to ask questions such as 'Are you ready to receive data?' or 'Have you any data for me?' Activities cannot be allowed to stop or start, or ask questions of another activity directly, for, as we have seen with the sharing of data, this would require that the activity giving the command or asking the question had knowledge of the state of the commanded or interrogated activity. Direct communication of activity control or interrogation signals would produce a method of synchronization which was not capable of verification. This is not to say that direct methods are not used: flags are very common and widely used. A flag is a variable which is held in common and can be accessed by several activities; the variable is used to indicate that a particular event has occurred. For example, the synchronization between the clock/calendar activity and the operator display activity in Example 5.1 is required at 5-second intervals; a flag in common could be set by clock/calender every 5 seconds and at intervals the operator display activity would check to see if the flag was set.

There are many problems with this method in large systems, some of which will be discussed in detail in later chapters. A particular disadvantage is that the operator activity has to continually check the flag and if any other activities are to take place, some means of enabling the operator display activity to wait before checking the flag again must be provided. What is required is a mechanism which enables the operator display activity to *wait* for an *event* (the 5-second time interval) to occur and for the clock/calendar activity to *signal* that the *event* has occurred.

If two procedures, WAIT(event) and SIGNAL(event), with the properties given below can be postulated then they can be used to describe all the necessary synchronization.

1. WAIT(event)   The task suspends activity as soon as the WAIT operation is executed and it remains suspended until the *event* is signalled as having occurred. If the *event* has already been signalled when the WAIT operation takes place the activity is resumed immediately (i.e., it simply continues without waiting).
2. SIGNAL (event)   The SIGNAL operation simply broadcasts the fact that the event has occurred and thus enables a waiting process to continue.

WAIT can be thought of as a procedure which continually reads the event status from a channel or a pool and SIGNAL as a procedure which writes information to a channel or a pool. Obviously if a single CPU is being used one activity cannot continuously read a particular variable – no other activity could change the status of the variable except for an interrupt activity – and hence for a WAIT procedure to work on a real machine some other method of implementing the procedure is required; at the design stage, however, we need not be concerned with how such a procedure can be implemented.

It should be noted that information about a particular event need not be limited to two activities, e.g., several activities may wait for the same event; however, normally only one activity would signal an event.

In a real-time system controlling plant, events on the plant may be significant and such events are SIGNALled by the use of external interrupts; these can be considered as synchronization signals. However, a requirement for correct operation of an external interrupt is that the activity being SIGNALled is

WAITing. Synchronization signals are represented as shown in Figure 5.20; note that hardware devices are shown as rectangular blocks.

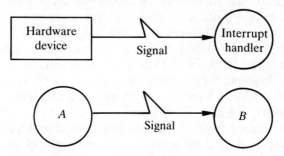

Fig. 5.20 Representation of signals.

5.9 EXAMPLE OF PRELIMINARY DESIGN

Once again we shall use the hot-air blower as our example. The outline structure is shown in Figure 5.4 and, using the notation described above, the system can be represented as shown in Figure 5.21. The representation is not unique, e.g., we have chosen to represent the plant interface as a pool and the display as an external unit connected by a channel to the operator display activity. An equally valid representation could be to treat the plant interface as two pools, one for input and the other for output, and also to treat the display as a pool. Treating the display as a pool would imply that information displayed on the screen could be read back. It should also be remembered that as the design proceeds the initial structure chosen may need to be modified.

The information passing either along channels or to and from pools is labelled as are the channels and pools themselves. Associated with the diagram is a table (Table 5.1) giving a brief description of each activity, pool, channel and signal.

In Table 5.1 the entries are not complete; as the design proceeds the definition of entries such as PLANT_DATA will have to be expanded to indicate what information about the plant is contained in it, e.g.

PLANT_DATA = SET_POINT;
      MEASURED_VARIABLE (* filtered value *);
      CURRENT_ERROR;
      CURRENT_OUTPUT;

These would then be further elaborated e.g.

SET_POINT = REAL RANGE 0–60, units °C, resolution 3 significant figures.

until at the end of the design process a detailed list of all activities, events and data would be available – this list is known as a *data dictionary*.

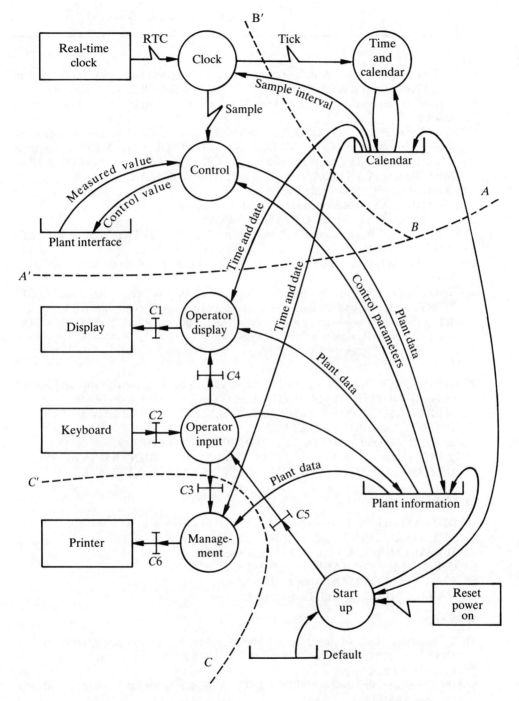

Fig. 5.21 Preliminary design of software for hot-air blower.

Table 5.1

ACTIVITIES

CLOCK: interrupt routine – handles external interrupt from real-time clock, sends signal to CALENDAR at TICK intervals and to CONTROL at SAMPLE_TIME intervals.

CALENDAR: activated by TICK, updates TIME and DATE which are held in CALENDAR pool.

CONTROL: activated by SAMPLE_INTERVAL, reads MEASURED_VALUE from plant, computes CONTROL_VALUE using CONTROL_PARAMETERS, outputs CONTROL_VALUE to plant and updates PLANT_DATA. CONTROL_PARAMETERS and PLANT_DATA are held in PLANT_INFORMATION pool.

OPERATOR DISPLAY: activated by
  (a) DISPLAY_TIME event in CALENDAR pool; or
  (b) an operator command UPDATE_DISPLAY
it updates the visual display unit with PLANT_INFORMATION and with OPERATOR_ACTION.

OPERATOR INPUT: activated by START_EVENT, reads and processes KEYBOARD_INPUT.

MANAGEMENT: activated by operator command which generates START_MANAGEMENT event, processes and controls the output of management information.

START: activated by power-up or reset, loads default values of PLANT_INFORMATION and CALENDAR, signals START_EVENT.

POOLS

PLANT INTERFACE: is the external interface and contains measured value and control value, i.e. OUTLET_TEMPERATURE and CONTROLLER_INPUT.

PLANT INFORMATION: contains CONTROLLER_PARAMETERS, SAMPLING_INTERVAL and PLANT_DATA.

CALENDAR: contains TIME, DATE, and DISPLAY_UPDATE data.

DEFAULT: contains default values of CONTROLLER_PARAMETERS, SAMPLING_INTERVAL, PLANT_DATA, TIME, DATE, and DISPLAY_UPDATE.

CHANNELS

C1: DISPLAY_OUTPUT – type CHAR.
C2: KEYBOARD_INPUT – type CHAR.
C3: MANAGEMENT_COMMAND – type MESSAGE.
C4: OPERATOR_COMMAND – type MESSAGE.
C5: START_UP_COMMAND – type MESSAGE.
C6: MANAGEMENT_OUTPUT – type MESSAGE.

SIGNALS

TICK: the basic unit of time in the system, must be an integer multiple of the CLOCK_INPUT_INTERVAL. SAMPLING_INTERVAL is expressed as an integer multiple of TICK.

SAMPLE: indicates that the SAMPLING_INTERVAL number of TICKs has occurred since previous SAMPLE.

Detailed Design: Module Subdivision

The preliminary design phase should reveal a number of questions relating to decisions which will have to be made as the design is refined. Examples are:

1. Will the keyboard generate an interrupt when a key is pressed?
2. Will the PLANT_INFORMATION be held in engineering units or in internal machine form or both? Where will conversions take place?
3. How is CLOCK to know if and when SAMPLING_INTERVAL is changed?
4. How is CONTROL to know when CONTROL_PARAMETERS or SET_POINT is changed?

5.10 DETAILED DESIGN: MODULE SUBDIVISION

The next stage in the design is to divide the system shown in Figure 5.21 into segments and to refine and subdivide the structure within each segment. The way in which segment division is carried out is largely dependent on the experience and skill of the designer, however, there are some simple guidelines:

1. Group activities with similar time-scales.
2. Group activities which are closely coupled to external plant.
3. Choose segments to minimize the number of channels crossing segment boundaries.

Application of Rule 3 leads to a division shown by the dotted line marked AA' in Figure 5.21. If we now apply Rule 2, the top part of the diagram can be divided as shown by the dotted line BB' and the application of Rule 1 leads to the division of the bottom half of the diagram as shown by dotted line CC'. In the following section the segment bounded by $A'BB'$, i.e., the segment containing the CLOCK and CONTROL activities, will be examined in more detail.

The first stage is to examine in greater detail the functions to be performed by the *activities* in the segment. These can be listed as follows:

1. CLOCK
   (a) service clock interrupt;
   (b) count clock interrupts to generate TICK;
   (c) count TICKs to generate SAMPLE; and
   (d) check if SAMPLE_INTERVAL has been changed.
2. CONTROL
   (a) read and process plant data;
   (b) calculate control value;
   (c) output control values to plant;
   (d) put PLANT_DATA into pool; and
   (e) check if SET_POINT or CONTROL_PARAMETERS have been changed.

There is probably little to be gained by attempting to subdivide the CLOCK activity any further.

Examination of the CONTROL activity, however, shows that only two of the actions – read plant data and output control values – are closely coupled to the plant and *require exact timing*. Based on this information an arrangement shown schematically in Figure 5.22 can be postulated and this is shown, using the design notation, in Figure 5.23.

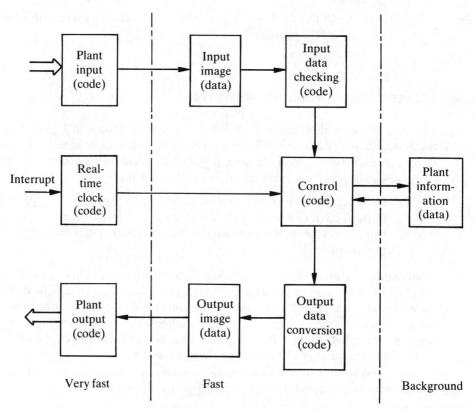

Fig. 5.22   Schematic for control software division.

At each sampling interval the real-time clock causes the plant inputs to be read and stored in some data area – referred to as the *input image* – and, either in parallel or immediately following, the control information, stored in a data area called *output image*, is sent to the plant. The information stored in the output image area is the controller values which were calculated on the basis of the previous input image data, i.e., the control output values always lag by one sample interval. The advantage of this approach is that it provides control outputs at accurately fixed sample intervals. The time (t_c) taken by the control calculation has only to satisfy the criterion $t_c < t_s$ and it does not matter if t_c varies with the data input providing that it remains less than t_s. The controller can be designed to compensate for the dead time introduced

Detailed Design: Module Subdivision

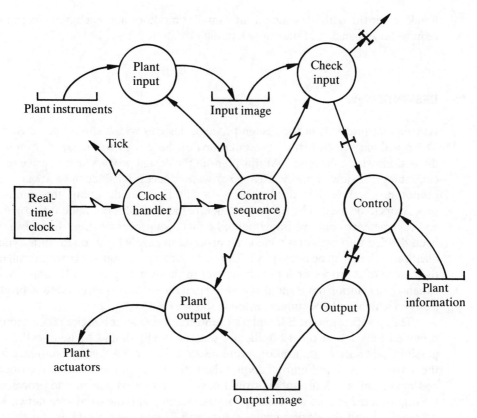

Fig. 5.23 Module division of control section.

through sampling by incorporating the dead time as part of the plant model. Examination of Figure 5.22 shows clearly how multiple processors could be used in real-time control since one obvious way of introducing parallel operations is to provide separate processors for PLANT INPUT, PLANT OUTPUT and CONTROL.

A further sub-division of the CONTROL activity has been made – into INPUT_CHECKING, CONTROL_CALCULATION and OUTPUT_CONVERSION. For a simple system this is probably unnecessary but, in general, plant input will have to be checked for alarm conditions, for instrument performance (drift, errors etc. [Iserman 1984]) and plant output will have to be modified to take into account particular actuator characteristics. It will be up to the designer to decide whether these operations are carried out as part of the PLANT_INPUT, PLANT_OUTPUT activities or in the separate activities INPUT_CHECKING and OUTPUT_CONVERSION shown in Figure 5.22. In the case of the hot-air blower, with only a single control loop, a much simpler arrangement is possible as is shown in Figure 5.23.

Each of the divisions of the system shown in Figure 5.21 needs to be refined to break down the activities shown into smaller modules and each new diagram will require the expansion of the data dictionary of Table 5.1.

5.11 DESIGN REVIEW

At frequent intervals in the design process, design reviews should be carried out. These will vary from formal presentations of the design to managers to informal discussions with colleagues. At this point in the design process an extensive review, combining a review of hardware and software is necessary since the software design cannot proceed further until a decision has been made about the number of processors to be used. The software structure has up to this point identified *activities* and each activity could be implemented on its own processor, i.e., the software has been divided into activities which can proceed in parallel (although some synchronization of them will be necessary). The costs of computer hardware would still make the choice of a processor for each activity in the system given in Example 5.1 too expensive a solution but it should be noted that the software design method produces a design which could be implemented in this way.

The specification for Example 5.1 requires that at least one processor should be provided for each of the 12 units and that each unit should be identical and two possible hardware configurations were given in Figure 5.4. Using Configuration 1, the activities given in Figure 5.23 and the activities required to cover the operator and management requirements would have to be carried out on one processor. If Configuration 2 is used a means of communicating with the local area network will be required and the design given in Figure 5.23 will need modifying as shown in Figure 5.25. The outcome of the review will be decisions which will influence particularly the communication mechanisms used between activities and the scheduling of activities. For example, a decision to use one processor for the activities shown in Figure 5.24 means that CONTROL_SEQUENCE, PLANT_INPUT, CONTROL_CALCULATION and PLANT_OUTPUT can be considered as one task which could be programmed as

```
PROCEDURE ControlTask;
  REPEAT
    Wait(tick);
    PlantInput;
    PlantOutput;
    ControlCalculation;
  UNTIL stop;
END ControlTask;
```

* * *

Example 5.3

If we assume that the specification of Example 5.1 is changed such that one hardware unit may be used to control all 12 plant units, then a design based on using one

Design Review

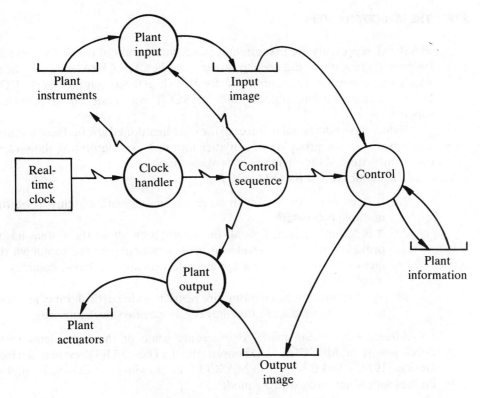

Fig. 5.24 Simplified module division.

processor for the data acquisition and actuator output, one processor for the control calculation and one processor for the operator/management communication can be considered. It should be noted that the three processors share memory to enable a pool for both the input and output images to be created.

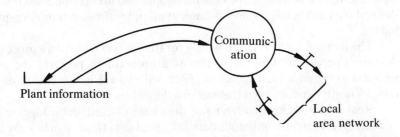

Fig. 5.25 Modification to provide for communication to a network.

5.12 THE MASCOT SYSTEM

MASCOT represents the first attempt at a formalization of the design of real-time systems. It arose out of the need to supplement the CORAL 66 language, the use of which was a mandatory requirement for British defence contract work. CORAL lacks any multi-tasking support and MASCOT was designed to provide such support.

It does have additional features to the ones mentioned above: these features are concerned with the grouping of activities into tasks and informing the underlying operating system of the characteristics of the tasks.

The major criticisms of MASCOT are:

1. It does not cope well with large systems or with systems to be run on multiple processors.
2. It is a 'flat' system, i.e., all the information about the system has to be provided at a single level and it does not support the top-down design methodology. There is a tendency to provide one large, complex, ACP diagram.
3. Synchronization mechanisms are restrictive; in particular it is not possible to synchronize tasks so that they run at specified clock intervals.

Attempts are being made to overcome some of these problems with the development of MASCOT 3 [Simpson 1984]. The DARTS system devised by Gomaa [1984, 1986] is similar to MASCOT but provides a wider range of abstract entities with which to construct a model.

5.13 STRUCTURED DEVELOPMENT FOR REAL-TIME SYSTEMS
[Ward and Mellor 1986]

A well established design method for non-real-time systems is the data-flow method, which views a program as a transformation of data. The basic notation used is shown in Figure 5.26 in which input data streams are transformed to give output data streams; the transformation is the equivalent of the activity in the MASCOT notation. It is assumed that the data being operated upon is time-discrete, that is, it is defined only at specific points in time; at all other times it is undefined or has null value.

The method has been used widely for transaction based data processing design. An example of this type of system is a program to perform the operation of withdrawals from a bank account. Each withdrawal is a separate discrete event (transaction); between such transactions the data is undefined.

Real-time systems involve more than time discrete data: logging and control systems involve time-continuous data and event data (logic signals). By time-continuous data we mean data that is defined on a continuous time spectrum. (Note that for design purposes the data, even after sampling, is treated as time-continuous.) In

Real-time Systems: Structured Development

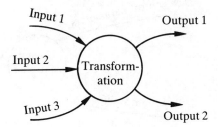

Fig. 5.26 Data flow notation.

order to deal with the different types of data and events, the data-flow notation has to be extended.

5.13.1 Data transformation

Examples of time-continuous transformations are shown in Figure 5.27. The continuous nature of the data is indicated by the use of the double arrow-head on the data-flow lines.

Time discrete transformation is shown in Figure 5.28. It should be noted that there is an ambiguity in Figure 5.28a. Should the diagram be interpreted to mean that the transformation can take place only if both inputs are present simultaneously; or if they arrive at different times will they be stored until both are present? To avoid this type of ambiguity the concept of synchronous transformation is used.

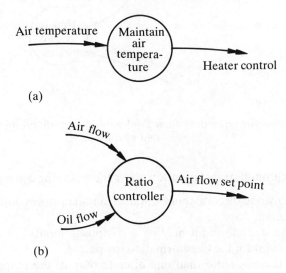

Fig. 5.27 Time-continuous data flow.

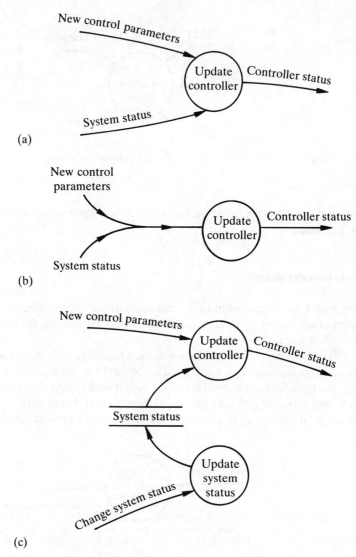

Fig. 5.28 Synchronous data flows. (a) Ambiguous diagram; (b) merged flows; (c) use of data store.

A synchronous transformation obeys the following conventions:

1. Only one discrete input (inputs from data stores and prompts – see below – are not counted.)
2. The discrete input may be a composite input, but all elements must be present for the transformation to operate.
3. If there is more than one discrete output, the outputs must be mutually exclusive.

Real-time Systems: Structured Development

Using these rules the transformation shown in Figure 5.28a can be represented either as in Figure 5.28b, in which the two inputs are combined to form a composite input, or as in figure 5.28c in which a data store is used to retain the system status.

A data store is represented by two parallel lines and it indicates that the data is retained between instances of time-discrete data. There are other means of combining data flows and these are summarized in Figure 5.29.

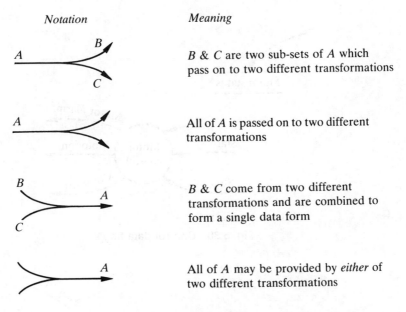

Fig. 5.29   Summary of data flow notation.

5.13.2 Control transformations

The notation used for control transformations is shown in Figure 5.30a. The event flows and transformations are denoted by dotted lines. It is implicit in the definition of a control transform that there is some internal memory and hence a control transform cannot simply pass through an event. Some external memory can be denoted. As an exact parallel to the data store, it is represented by two parallel dotted lines and is referred to as an event store; this is shown in Figure 5.30b.

5.13.3 Prompts

In order to clearly show the interrelationship between event flows and data flows the concept of a prompt is used. The three prompts are ENABLE, DISABLE and TRIGGER and their use is illustrated in Figure 5.31. Prompts are not considered as inputs to a transformation but simply as switches which are used to turn a

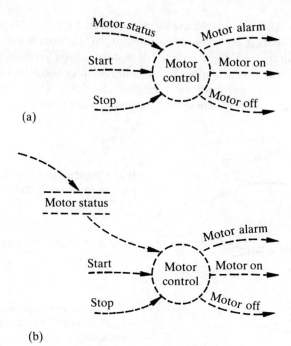

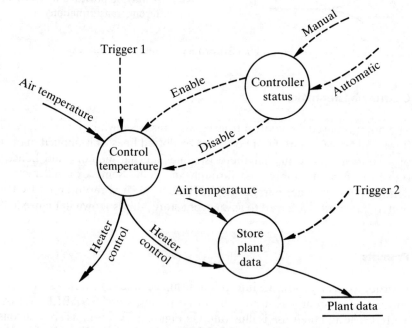

Fig. 5.30 Control data flows.

Fig. 5.31 Data flow showing use of prompts.

transformation on or off. The TRIGGER prompt causes the transformation to operate once and can be used to transform time-continuous data into time-discrete data.

5.13.4 Summary of the method

The system provides a constrained graphical notation for representing the general structure of a real-time system. The rules, summarized in Table 5.2, provide a simple check of consistency of the interfaces between the various transformations. Because the data flows are shown clearly it becomes easy to apply the design heuristic to minimize the data flowing across boundaries in deciding how to divide the system into modules. (In doing this it should be remembered that the notation allows for composite data flows and hence a single line may represent a complex data structure and hence a complex interface.)

Table 5.2 Input/output summary for transformations

Transform	Inputs	Outputs
Data	Data flow	Data flow
	Prompts	Event flow
Control	Event flow	Event flow
	Prompt	Prompt

5.13.5 Building the model

The method requires two models to be built, an abstract model and an implementation model. The models will normally be constructed in parallel since high-level implementation details may affect lower-level decisions for the abstract model. The phases involved in building the abstract model are illustrated in Figure 5.32.

The first stage is to define the system in terms of its connections with the environment. This is done by means of a context schema and an event list. To illustrate this a simple system consisting of the control loop for the temperature control of the hot-air blower is considered. It is assumed that an operator can enter a new set point for the temperature and that an alarm is provided which is turned on if the temperature goes outside preset limits. The context schema for this system is shown in Figure 5.33. The event list for the system is very simple; it consists of one entry:

> Operator changes temperature set point.

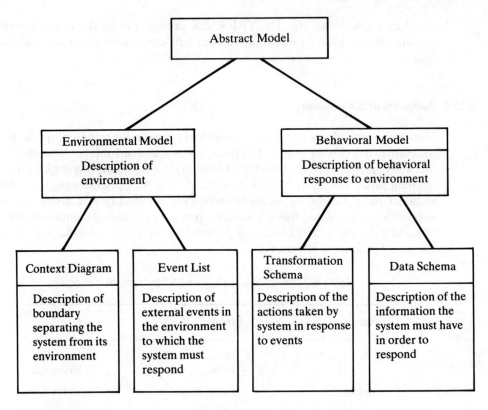

Fig. 5.32 Outline of abstract modelling approach of Ward and Mellor.

At this stage in the design a data dictionary would also be started. From the context schema the following entries can be made:

air-temperature = * temperature of the air at the tube outlet values: 0..100; units: degrees centigrade *
heat-demand = * control signal to heater: values: 0..10; units: volts *
alarm-signal = * operator alarm : values: on/off *

The signals indicated on the context schema as coming from the environment are not necessarily the direct outputs from transducers on the plant, or from keyboards, or displayed. In fact, the method assumes that some processing of the data will take place in order that what is presented to the system indicated in the context schema will be an image of the environment, rather like the plant input and plant output image discussed earlier. For example, the temperature transducer indicated in the terminator box in Figure 5.33 is a virtual transducer. It is made up of the actual transducer and some software as is shown in Figure 5.34a and 5.34b. The figures include some implementation details: in Figures 5.34a there is the statement

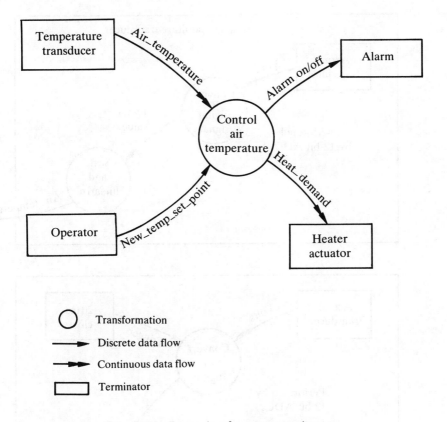

Fig. 5.33 Example of a context schema.

that the analog-to-digital conversion is to be performed using a 12-bit converter and in figure 5.34b the information that the conversion is to take place every 20 ms is added. It should be noted that the output from the 'convert to digital signal' transformation is a discrete data value and that the 'scale and linearize' transformation has to convert it back to a continuous value. The reason for converting back (or showing it converted back) is to avoid implying at this stage in the design that there is any synchronization between activities in the virtual temperature transducer and the main software system. The air temperature being made available as a continuous signal implies that the main system may sample it at any time.

The second stage of the modelling is to develop the behavioral model. This is the model which shows how the system responds to activities taking place in the environment. The model can be developed in two ways: transformation schema or data schema. The transformation schema describes the actions which have to be taken in response to events in the environment; the data schema provides a description of the data which the system must have in order to respond. In the

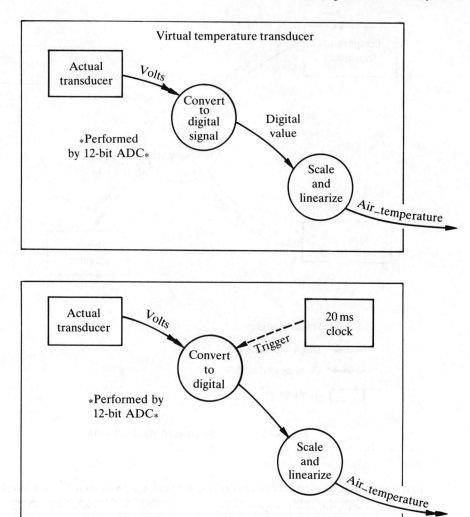

Fig. 5.34 Interface transducer modelling.

example being considered the transformation schema is the most appropriate method to use and the schema is shown in Figure 5.35.

Each of the transformation bubbles is numbered and further sets of diagrams could be drawn showing an expansion of each of the transformations until a level is reached at which the operations to be performed within each transformation cannot be subdivided further. In this example there is no need to go further than this first level of diagram. A *transformation specification* corresponding to each transformation on the diagram is drawn up.

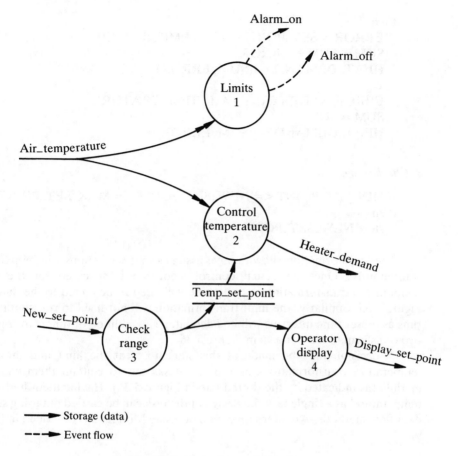

Fig. 5.35 Transformation schema.

1. Limits

Continuously
If AIR_TEMPERATURE <= MAXIMUM_TEMPERATURE
and AIR_TEMPERATURE >= MINIMUM_TEMPERATURE
then issue ALARM_OFF

If AIR_TEMPERATURE > MAXIMUM_TEMPERATURE
and AIR_TEMPERATURE < MINIMUM_TEMPERATURE
then issue ALARM_ON

2. Control temperature

Continuously
If MIN_HEAT_DEMAND <= HEAT_DEMAND <= MAX_HEAT_DEMAND

then
   ERROR = SET_POINT - AIR_TEMPERATURE
   SUM = SUM + ERROR
   HEAT_DEMAND = f(SUM, ERROR)
else
   ERROR = SET_POINT - AIR_TEMPERATURE
   SUM = SUM
   HEAT_DEMAND = f(SUM, ERROR)

3. Check range

If MIN_SET_POINT <= NEW_SET_POINT <= MAX_SET_POINT
then
   store NEW_SET_POINT

The transformation shown above assumes that the actions will be performed continuously. Once an actual control algorithm is specified for the control temperature transformation, a cycle time will need to be added to the diagram in Figure 5.35. Similarly, the limit transformation will not need to run continuously: runs as preset time intervals will be adequate. The addition of timers to trigger the two transformations is shown in Figure 5.36.

A decision may be made at this stage to treat the limit and the control temperature transformations as separate tasks but to contain them within one module (as indicated by the dotted line in Figure 5.36). Having identified 'control temperature' as a single task the design of the task can be carried out using standard data-flow methods. An alternative method using Modula-2 is described in Chapter 9.

5.14 SUMMARY

In this chapter we have discussed three approaches to the software design of real-time systems: single program, foreground/background and multi-tasking. Of these methods, the one based on multi-tasking is the most general. It has the advantage that, through the use of abstract entities (pools, channels, signals and activities in the MASCOT notation; or data and event flows with data and control transformations in the data flow notation), the designer is allowed to concentrate on the application aspects of the problem. At a later stage of the design the various activities identified can be coalesced into either the single program or foreground/background methods if they will provide the most appropriate solution.

The method also separates the initial software design from dependence on a particular hardware solution. The internal structure of activities can be designed as if they were sequential programs. The actual implementation details of pools, channels, signals, activities data and control transformations becomes a separate

Summary

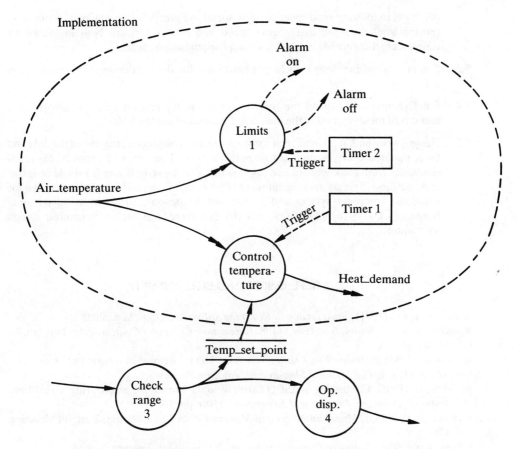

Fig. 5.36 Transformation schema showing the addition of task timing.

problem. Activities (or transformations) can be designed as if they were sequential programs using the standard techniques (see, e.g., Pressman 1983) and we will not consider these methods. In subsequent chapters we will consider how tasks can be created and how the communication between tasks can be handled, both by using conventional operating systems and by the use of modern real-time languages.

EXERCISES

5.1 Criticize the requirements specification given in Example 5.1. What information is missing? Rewrite the specification to include the missing information.

5.2 Write a requirements specification for an automobile instrument display system. The display is to indicate road speed, engine speed, oil pressure (normal/low), water level (normal/low). Assume that engine speed and road speed are both measured by transducers that provide an analog signal proportional to speed.

5.3 Prepare an outline design, including hardware, for the system described in question 5.2.

5.4 For Example 5.1, expand the operator input/display and management information sections of the design using the methods described in section 5.10.

5.5 Design, write and, if possible, test a system which (a) displays on the screen the date and time, updated every second; (b) rings the keyboard bell every 5 seconds; (c) reads characters from the keyboard and displays them on the screen. Use (a) single program approach and (b) multi-tasking approach. (Note: the system will have to be designed for a specific computer system and it may not be possible to do it on all personal computers.) Use (a) MASCOT and (b) the Ward and Mellor structured design technique.

REFERENCES AND BIBLIOGRAPHY

ALLWORTH, S.T. (1981), *Introduction to Real-time Software Design*, Macmillan
BENNETT, S. and LINKENS, D.A. (eds.) (1984), *Real-time Computer Control*, Peter Peregrinus, Stevenage
BOOCH, G. (1983), *Software Engineering with Ada*, The Benjamin/Cummings Pub. Co.
BROOKS, F. (1975), *The Mythical Man-month*, Addison Wesley
BUDGEN, D. (1985) 'Combining MASCOT with Modula-2 to aid the engineering of real-time systems', *Software: Practice and Experience*, **15(8)**, pp. 767–93
CASSELL, D.A. (1983), *Microcomputers and Modern Control Engineering*, Reston Publishing Co
COATS, R.B. (1982), *Software Engineering for Small Computers*, Edward Arnold
FOX, J.M. (1982), *Software and its Development*, Prentice Hall
GAINES, B.R. and SHAW, M.L.G. (1984), *The Art of Computer Conversation*, Prentice Hall
GLASS, R.L. (1982), *Modern Programming Practices: a Report from Industry*, Prentice Hall
GOMAA, H. (1984), 'A software design method for real-time systems', *Communications ACM*, **27(9)**, pp. 938–49
GOMAA, H. (1986), 'Software development of real-time systems', *Communications ACM*, **29**, p. 657
HATLEY, D.J. (1984), 'A structured analysis method for large real-time systems', *The Heap, DEC SIG Structured Languages*, **7(2)**, pp. 21–38
ISERMAN, R. (1984), 'Process fault detection based on modelling and estimation methods – a survey, *Automatica* **20(4)**, pp. 387–404
JACKSON, K. and SIMPSON, H.R. (1975), 'MASCOT – A modular approach to software construction, operation and test', *RRE Tech. Note*, No. 778
JACKSON, M. (1983), *System Development*, Prentice Hall
JOHNSON, C.D. (1984), *Microprocessor-based process control*, Prentice Hall
MELLICHAMP, D.A. (ed.) (1983), *Real-time Computing with Applications to Data Acquisition and Control*, Van Nostrand Reinhold
MYERS, G.J. (1978), *Composite Structured Design*, Van Nostrand Reinhold

PARNAS, D.L. (1972), 'On the criteria to be used in decomposing systems into modules', *Communications ACM* **15(12)**, pp. 1053–8

PARNAS, D.L., CLEMENTS, P.C. and WEISS, D.M. (1985), 'The modular structure of complex systems', *IEEE, Software Engineering*, **SE11(3)**, pp. 259–66

POMBERGER, G. (1986), *Software Engineering and Modula-2*, Prentice Hall

PRESSMAN, R.S. (1982), *Software Engineering: a Practitioner's Approach*, McGraw-Hill

SHOOMAN, M.L. (1983), *Software Engineering*, McGraw-Hill

SIMPSON, H.R. and JACKSON, K. (1979), 'Process synchronization in MASCOT, *Computer Journal* **22(4)**, pp. 332–45

SIMPSON, H.R. (1984), MASCOT 3, *IEE Colloquium*, Paper No. 3

SOMMERVILLE, I. (1982), *Software Engineering*, Addison Wesley

STEVENS, W.P., MYERS, G.J. and CONSTANTINE, L.L. (1974), 'Structured design', *IBM Systems J.* **13**, p. 15; also in Yourdon, E.N. (1979), *Classics in Software Engineering*, Yourdon Press

TZAFESTAS, S.G. (ed.) (1983), *Microprocessors in Signal Processing, Measurement and Control*, Reidel, Dordrecht

WARD, P.T. and MELLOR, S.J. (1986), *Structured Development for Real-time Systems*: Vol. 1 *Introduction and Tools*; Vol. 2 *Essential Modelling Techniques*, Yourdon

ZAVE, P. (1980), 'The operational approach to requirements specification for embedded systems', Report, December, University of Maryland, College Park

ZAVE, P. (1982), 'An operational approach to specification for embedded systems', *IEEE Trans. Software Engineering*, **SE8(3)**, pp. 250–69

ZAVE, P. (1984a), 'The operational versus conventional approach to software development', *Communications ACM* **27(2)**, pp. 104–18

ZAVE, P. (1984b), 'An overview of the PAISLey project – 1984', Tech. Rep., AT and T Bell Laboratories

ZAVE, P. (1984c), 'The anatomy of a process control system'. Tech. Rep., AT and T Bell Laboratories

6

Operating Systems

6.1 INTRODUCTION

The problem of design is simplified if details of the lower levels of implementation on a specific computer using a particular language can be hidden from the designer. An example of this approach was given in the previous chapter when it was assumed that the system was being designed to run on a MASCOT machine. This was one example of the use of a virtual machine. It implies that some software will be provided to convert the hardware of an actual computer, or the hardware plus an existing operating system, in such a manner that it appears to behave like a MASCOT system [Budgen 1985].

The provision of a MASCOT virtual machine is an extension of an approach that has been used for many years, in that an operating system for a given computer converts the hardware of the system into a virtual machine with characteristics defined by the operating system. Operating systems were developed, as their name implies, to assist the operator in running a batch-processing computer; they then developed to support both real-time systems and multi-access on-line systems.

The traditional approach to providing a system to meet a wide range of requirements has been to incorporate all the requirements inside a general purpose operating system. The typical arrangement is illustrated in Figure 6.1. Access to the hardware of the system and to the I/O devices is through the operating system. In many real-time and multi-programming systems restriction of access is enforced by hardware and software traps. The operating system is constructed, in these cases, as a monolithic monitor. In single-job operating systems access through the operating system is not usually enforced, but it is good programming practice to do so; additionally, it facilitates portability in that the operating system entry points will not be changed in different implementations. In any general purpose system there will be some facilities which will be required in a particular application. The traditional approach is to provide a mechanism which allows a limited range of modifications; this process is known as system generation (SYSGEN).

In addition to supporting and controlling the basic activities, operating systems typically provide various utility programs, e.g., loaders, linkers, assemblers and debuggers, as well as run-time support for high-level languages.

In recent years there has been considerable interest in developing an approach in which only a minimum kernel or nucleus of the operating system is provided on to which additional features can be added by the applications programmer writing

Introduction

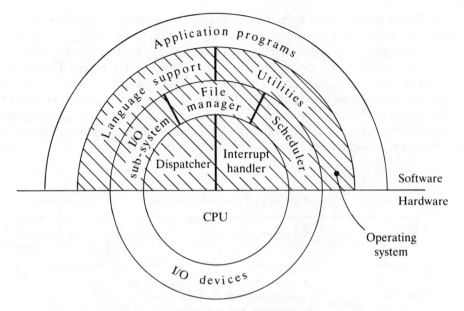

Fig. 6.1 General purpose operating system.

in a high level language. This structure is shown in Figure 6.2; it is only the nucleus or kernel to which access is limited. In this type of system the distinction between the operating system and the application software becomes blurred; it enables all the operating system functions to be modified or designed to suit the application. This type of approach is considered in Chapter 7.

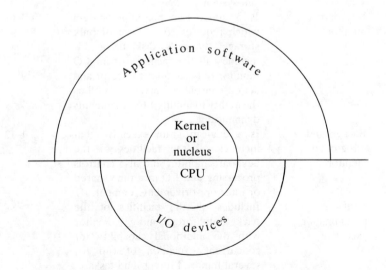

Fig. 6.2 Minimal operating system.

There are many different types of operating system; the position has grown more complicated with the development of ROM-based '*monitors*' which are frequently provided by manufacturers. These monitors have grown out of bootstrap loaders and now provide additional support features for terminals and simple input/output operations. The various types of monitors/operating systems have been categorized by Mellichamp [1983, p. 347] and are summarized in Table 6.1. The major emphasis in monitors, as in the single-user, single-job operating system described below, is in communication with the user and in providing simple support for the input/output devices.

Table 6.1 Classifications of operating systems (from Mellichamp)

System	*Key characteristics*	*Memory size (K words)*
Terminal interface monitor	Is used only by the simplest single application dedicated computers. It will handle process interrupts and allow the user to start a program, display or alter memory locations, set breakpoints and load or dump programs.	2
Input/output monitor	Handles its own initialization and control as well as communication between itself, system programs, user programs, and a simple I/O sub-system.	2
Keyboard monitor	Is a single-user operating system which requires some form of bulk storage, typically disk drives. It includes all the facilities of an I/O monitor plus routines to accept and act on console commands as well as the ability to modify I/O assignments dynamically.	3–6
Background/ foreground monitor	Is a dual program executive that includes all the facilities of the keyboard monitor and also controls processing and I/O in a time-shared or interrupt-driven environment.	4–12
Multi-programming executive	Includes all the facilities of the background/foreground monitor with the additional capability for concurrent execution of (up to) several hundred foreground tasks.	>8

In the following sections operating systems of the following types will be considered:

- single-user, single-job;
- foreground/background; and
- multi-tasking.

The emphasis will be on the use of such systems for real-time control applications.

6.2 SINGLE-TASK OR SINGLE-JOB OPERATING SYSTEM

As an example of a single-user, single-task, disk-based operating system, the CP/M 80 system of Digital Research will be described. This system is available for 8080 and Z80 based computer systems.

CP/M consists of three major sections:

- console command processor (CCP);
- basic input/output system (BIOS); and
- basic disk operating system (BDOS).

The relationship between the various sections of the operating system, the computer hardware and the user is illustrated in Figure 6.3.

The console command processor provides a means by which the user can communicate with the operating system from the computer console device. It is used to issue commands to the operating system and to provide the user with information about the actions being performed by the operating system. The actual processing of the commands issued by the user is done by the BDOS which also handles the input and output and the file operations on the disks. The BDOS makes the actual management of the file and input/output operations transparent to the user. Application programs will normally communicate with the hardware of the system through 'system calls' which are processed by the BDOS.

The BIOS contains the various device drivers which manipulate the physical devices; this section of the operating system may vary from implementation to implementation as it has to operate directly with the underlying hardware of the computer. For example, depending on the manufacturer the physical addresses of the peripherals may vary, as may the type of peripheral, and the type of controller used for the disk drive. All these differences will be accommodated in the coding of the BIOS.

On starting the computer the three sub-systems which reside on the disk are loaded into the memory. The computer must, therefore, have some means of booting the operating system from a systems disk. The normal arrangement is to provide a small bootstrap loader in a ROM chip. Following system boot the memory allocation is as shown in Figure 6.4. The combination of BDOS and BIOS is frequently referred to as the FDOS (functional disk operating system) and these two units have to remain in memory. Since the CCP functions cannot be called from a

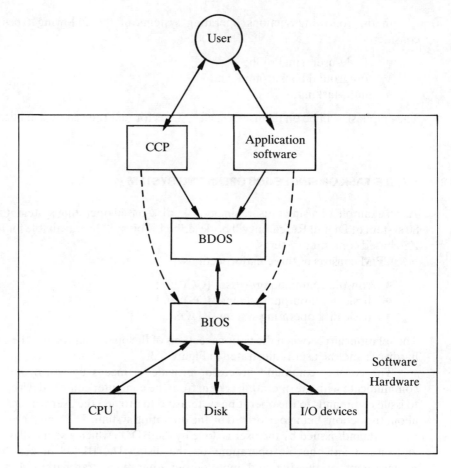

Fig. 6.3 General structure of CP/M.

user program, the CCP area can be over-written by a user program provided that CP/M is reloaded when the program terminates.

6.2.1 CCP direct commands

There are seven direct commands available to the user; the routines to support these commands are held in the CCP area of memory. Full details of the commands are given in the manual *Introduction to CP/M Features and Facilities* issued by Digital Research and in the various guides to the use of CP/M. The commands are listed briefly below:

1. DIR ⟨afn⟩   Display a list of all filenames which satisfy the ambiguous

Single-task or Single-job Operating System

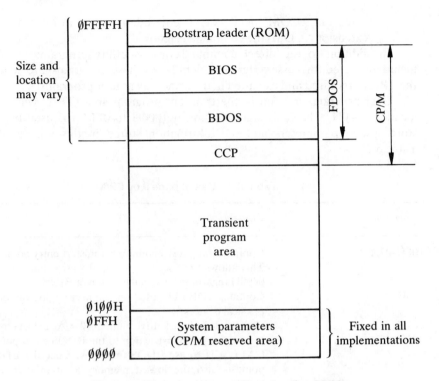

Fig. 6.4 Typical memory allocation for CP/M.

filename 'afn'. If the parameter ⟨afn⟩ is not supplied all the filenames on the current drive are listed. Various switches can be added to the parameter to display, e.g., the size of each file.

2. ERA ⟨afn⟩   All files which satisfy the ambiguous filename 'afn' are erased. All the files on the disk can be erased by setting ⟨afn⟩ equal to *.*.
3. REN ⟨new⟩ = ⟨old⟩   If the file ⟨old⟩ exists and there is no file with the name ⟨new⟩ then the file ⟨old⟩ is renamed as ⟨new⟩. If a file with the name ⟨new⟩ exists then the command is aborted. If the file ⟨old⟩ does not exist then the filename is displayed on the console followed by a question mark.
4. SAVE ⟨n⟩ ⟨ufn⟩   Starting at the address ∅1∅∅H, n blocks of 256 bytes of memory are written to the unambiguous filename 'ufn'. Note if the file ⟨ufn⟩ already exists, the contents of the file will be overwritten.
5. TYPE ⟨ufn⟩   The contents of the file ⟨ufn⟩ are listed on the console.
6. USER ⟨n⟩   This is a facility to enable separate file areas to be maintained on the disk. Any integer value in the range ∅ to 15 can be entered.
7. ⟨ufn⟩   Entering an unambiguous filename of a file which has the extension '.COM' is interpreted as requesting the file to be loaded into memory and executed. The file will be loaded into the TPA and run. Note the file

extension '.COM' need not be entered. If the ⟨ufn⟩ does not have an extension '.COM' the command is ignored.

In addition to the direct command various utility programs are provided including an editor, assembler and debugger; these programs are loaded into the TPA as required and executed in the same way as user programs.

User programs are run in the transient program area (TPA) which starts at location Ø1ØØH. The first page of memory (ØØØH to ØFFH) is used by CP/M to store system parameters and the information stored in this area is shown in Table 6.2.

Table 6.2 Use of page Ø by CP/M

Location	Contents
ØØH–Ø2H	Contains jump instruction to warm start entry point in BIOS. This allows the use of a standard warm start entry of JMP ØØØØH regardless of actual location of BIOS.
Ø3H	Contains IOBYTE used in connecting logical devices to physical devices.
Ø4H	Current default disk drive number (Ø = A, 1 = B, etc.).
Ø5H–Ø7H	Contains a jump instruction to the BDOS entry point, hence CALL Ø5H to use BDOS functions. Since the BDOS entry point is also the lowest memory location used by CP/M locations Ø6H and Ø7H give the lowest memory address used by CP/M.
Ø8H–27H	Standard interrupt locations – not used by CP/M.
3ØH–37H	Interrupt location 6, not currently used but reserved for future use by CP/M.
38H–3AH	Entry point to the debugger from programmed breakpoints.
3BH–3FH	Reserved for future use.
4ØH–4FH	Scratch area for customized BIOS.
5ØH–5BH	Reserved for future use.
5CH–7CH	Default file control block, produced for a transient program by the console command processor.
7DH–7FH	Reserved for future use.

6.2.2 Basic disk operating system

The CCP facilities are useful in developing user programs, but one of the purposes of the operating system is to support user programs while they are running. The facilities provided by the BDOS are used for this purpose and these are essentially concerned with operating the input and output devices attached to the system.

Input/output devices

Two types of input/output devices are supported:

1. Sequential. Data is transferred one byte at a time,
2. Blocked. Data is transferred in fixed length blocks or records of 128 bytes: these are assumed to be disk drives in CP/M.

In dealing with input/output devices CP/M considers devices to be 'logical' or 'physical'. Logical devices are software constructs used to simplify the user interface: user programs perform input and output to logical devices, the BDOS connects the logical device to the physical device. The actual operation of the physical device is performed by software in the BIOS.

Sequential devices

CP/M supports four sequential logical devices, for historical reasons they are named CON:, RDR:, PUN: and LIST:. (Note that the colons are part of the names.) They have the following characteristics:

1. CON: This is the 'console' device used to communicate with the user.
2. RDR: This is a general purpose serial input device which was originally intended for a paper-tape reader.
3. PUN: This is a general purpose serial output device which was originally intended for a paper-tape punch.
4. LST: This designates the listing device. It is normally used to direct output to a printer, but can be used to send information to a mass storage device other than a disk.

Associated with each logical device is one or more logical device drivers; these are:

1. CON: (a) CONIN Input one character at a time from the console input device.
    (b) CONOUT Output one character at a time to the console output device.
    (c) CONST Examine the console input device to test if an input (a character) is pending.
2. RDR: READER Read one character from the serial input device.
3. PUN: PUNCH Output one character to the serial output device.
4. LST: (a) LIST outputs one character at a time to the list output device.
    (b) LISTST tests the list device to determine if it is ready to receive a character or if it is still busy.

CP/M makes provision to map the logical devices onto 12 physical devices; the actual mapping is controlled by the bit pattern in the IOBYTE which is stored in location ØØØ3H in the memory. The arrangement of logical and physical devices is shown in Figure 6.5. Illustrated is the CON: device which is assumed to be connected to a VDU and a keyboard. It should be noted that it is only at the BIOS level that

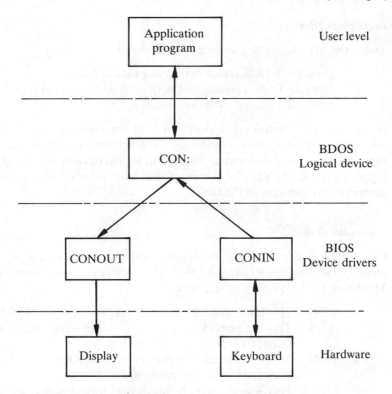

Fig. 6.5 Logical and physical I/O devices.

information regarding the physical characteristics of the device, e.g., address on IO bus, data-transfer rates, is required. At the logical device level all that is required is to know, for example, that CON: is a sequential device capable of input and output.

Blocked devices

Up to 16 physical devices which use blocked format can be supported and CP/M assumes that they will be disk drives. Only one logical driver is supported. This driver assumes that data is stored in blocks of 128 bytes; if the physical device uses a different block size, the physical device driver in the BIOS has to split the physical blocks into 128-byte logical blocks (or if the block size is smaller, group the blocks).

Associated with the driver is an FCB (file control block) which is located in the first page of memory and holds information about the file which is to be accessed by the driver. The format of the FCB is shown in Table 6.3.

It is the user's responsibility to put the correct information into the FCB before using any of the disk operations available in the BDOS. Note that if several files are to be accessed by the user program, the program must provide storage for each

Single-task or Single-job Operating System

Table 6.3 File control block format for CP/M

	0	1	2	3	4	5	6	7	8	9	A	B	C	D	E	F
ØØ	DN	FN								FT			EX	Ø	Ø	RC
1Ø	DM															
2Ø	CR	RN														

DN	Disk drive reference – values are in the range Ø to 16 where Ø refers to the current drive (i.e., the drive whose number is stored in location ØØØ4H) and 1 to drive A, 2 to drive B etc.
FN	These locations contain the first eight characters of the file name, the entry is left justified and filled with spaces.
FT	The three characters of the file name extension are stored in these locations. The high bits of bytes Ø9 and ØAH are used to specify the attributes of the file. Normally these bits are set to Ø: if the high bit of byte Ø9 is set to 1 then the file is read only; if the high bit of byte ØAH is set to 1 then the file is a system file which cannot be listed by the DIR command.
EX, RC and DM	These are used by BDOS to hold information about the location of the file on the disk and should not be changed by the user.
CR	This contains the current record to be accessed by the driver; it is normally set to Ø by the user on entry in order to start reading from the beginning of the file.
RN	This is optional; if used it contains the number of a record which is to be read, i.e., it provides for random access to records.

FCB; if only one file is to be used then the FCB area in locations ØØ5CH to ØØ7CH can be used (this is the default area used by the console command processor).

System calls

Both the sequential and blocked operations provided by the BDOS are called in the same way. Locations ØØØ5H to ØØØ7H contain a jump instruction to the start of the BDOS and the appropriate operation is selected by loading the C register with a number corresponding to the operation required and using an instruction 'CALL 5'; any parameters required are passed in the various CPU registers. By using this approach the BDOS can be located in different areas of memory for different implementations of CP/M and yet application programs which use the facilities solely through calls to the BDOS will run on the different systems.

* * *

Example 6.1 Use of system calls in CP/M

The assembly coding given below shows the coding required to transfer a character to the CON: output.

```
; Example on the use of BDOS calls
;
; Define symbols
;
BDOS        EQU    5       ; address of entry point for BDOS system calls
WRCH        EQU    2       ; function number for CON: output
;
; output character 'X'
;
            LD E,'X'         ;parameter to be passed
            LD C,WRCH        ;BDOS function number
            CALL BDOS        ;call BDOS
;
```

* * *

The various functions which are available in a typical BDOS for CP/M are listed in Table 6.4; the actual number of functions provided will vary with the version number of CP/M being used, but it should be noted that there is compatibility of function number between the different versions of CP/M.

In addition to the BDOS routines listed in Table 6.4 the user can also gain access

Table 6.4   System (function) calls to BDOS

Function number	Name	Action
0	RESET	Re-initializes the disk system.
1	RDCH	Waits for ASCII character to be typed at console and returns it in A.
2	WRCH	Sends character in E to console and waits until character has been accepted.
3	TAPIN	Waits for character from READER, returns it in A.
4	TAPOUT	Sends character in E to logical PUNCH device.
5	PRNTER	Sends character in E to logical LIST device.
6	RAWIO	If on entry E contains $\emptyset$FFH then this is treated as an input request and the console input register is examined; if a character is present it is returned in A; if no character is present then the value $\emptyset$H is returned. If the value in E is not $\emptyset$FFH then the character in E is output. This routine is used to get and send the full ASCII character set; functions 1 and 2 trap certain control characters and interpret them.
7	GETIO	IOBYTE value is returned in A.
8	SETIO	Value in E is transferred to IOBYTE.
9	PRSTR	Sends a string character by character to console until $ is encountered; the address of the first character in the string is held in DE.
10	RDLINE	Reads characters from console and places them in buffer with address in DE until CR is entered.

Table 6.4 System (function) calls to BDOS (continued)

Function number	Name	Action
11	CSTAT	Returns a ØH in register A if no character is available at the console; it returns non-zero if character is available.
12	VERNUM	Returns the version number of CP/M in registers HL.
13	RSTDSK	All disk drives are reset.
14	SETDSK	Selects the disk drive whose number is in E.
15	OPEN	Must be used before attempting to access an existing file; DE contains a pointer to the FCB.
16	CLOSE	Close an existing file; DE points to FCB.
17	SRCHF	Searches for a file with the ambiguous file name contained in the FCB. DE contains pointer to FCB. The search starts at the beginning of the directory.
18	SRCHN	Searches for a file as in SRCHF, but search begins at previous match.
19	DEL	Delete file; DE contains pointer to FCB.
20	RDSEC	Read sector into current disk buffer (default buffer starts at location ØØ8ØH and is 128 bytes long); DE contains pointer to FCB.
21	WRSEC	Write sector (128 bytes) from current buffer to disk; DE points to FCB.
22	CREAT	Create new file, DE points to FCB.
23	REN	Rename file, old name is in bytes 1–11 of FCB, new name is in bytes 16–26 of FCB.
24	LOGV	Returns in HL information regarding which drives are online and which are offline.
25	CURD	The current disk drive number is returned in A.
26	CURB	Sets the current disk buffer to the address contained in DE.
27	ALLOC	Returns information in HL about the allocation vector for the current disk.
28	WRPR	Sets current disk in write protect mode.
29	GETR	Returns information about which drives are currently set to read only.
30	SATT	Sets file attributes.
31	DPAR	Returns address of disk parameter block.
32	SUSR	Allows display and setting of current user number.
33	RDRN	Read a random record from file to disk buffer.
34	WRRN	Write a random record.
35	CFS	Return number of records in a file.
36	SRR	Returns current random record number.
37	RSDR	Reset several drives as specified in value in DE.
40	WRZ	Similar to WRRN except that the first time a previously unallocated group is written all records in the group except the one being written are set to ØØ.

Table 6.5 BIOS entry points.

```
;
; BIOS jump table
;
BIOS:       JP    CBOOT      ; cold boot entry point
WBOOTE:     JP    WBOOT      ; warm boot entry point
            JP    CONST      ; console status
            JP    CONIN      ; console input
            JP    CONOUT     ; console output
            JP    LIST       ; list device output
            JP    PUNCH      ; punch device output
            JP    READER     ; reader device input
            JP    HOME       ; restore disk drive
            JP    SELDSK     ; select disk drive
            JP    SETTRK     ; select track number
            JP    SETSEC     ; select sector number
            JP    SETDMA     ; select buffer address
            JP    READ       ; read sector
            JP    WRITE      ; write sector
            JP    LISTST     ; list device status
            JP    SECTRAN    ; translate skewed sector
                             ; number
```

to the BIOS routines, since the start of the BIOS is stored in location ∅∅∅∅H and the first locations in the BIOS contain a jump table to the various routines. The standard BIOS entry points are listed in Table 6.5.

Although it is not a real-time multi-tasking operating system the CP/M system has been dealt with at some length since it illustrates some of the features of a traditional operating system. The emphasis is on support for the disk file system and the input/output devices. Access to the system functions is by means of sub-routine calls and information is passed in the CPU registers of the machine. Because of the latter, the functions cannot be called directly from most high-level languages and hence there is a degree of isolation between the operating system and a programmer using a high-level language. The isolation is deliberate: it is an example of 'information hiding'. The connection between the high-level language and CP/M is made by the compiler writer through the provision of run-time support routines which convert the CP/M virtual machine into the virtual machine described by the high-level language.

The isolation is not complete in that it is possible to call assembly-coded routines from high-level languages and to pass parameters between the high-level language code and the assembly code; this does require some detailed knowledge of the system, however. Again, information hiding is used in that the details of the physical implementation on the CPU and of the IO devices are hidden within CP/M and hence operations are performed on the CP/M virtual machine.

6.3 SIMPLE FOREGROUND/BACKGROUND OPERATING SYSTEM

It is possible to convert CP/M based machines into a simple foreground/background system. In order to do this some form of real-time clock must be available on the computer – a suitable clock can be set up by using the Zilog CTC chip – and it must be possible to connect the clock chip to the IRQ interrupt line on the CPU chip. In most CP/M systems the interrupts are not used, hence the clock interrupt will be the only one on the system. If there are other interrupts, the CTC chip must be connected to the interrupt daisy chain. (Some systems provide a 'type-ahead' facility which uses interrupts from the keyboard to input characters to a buffer.)

Caution: the successful conversion of CP/M to a foreground/background system using interrupts depends on how the BIOS has been implemented for a given system. In particular it may not be possible to use the disk file system if the BIOS does not protect critical disk operations, i.e., sector reading and writing, by disabling interrupts during such operations.

The TPA part of the memory can be divided into two parts: one part will contain the program which will be entered when an interrupt occurs – the foreground – and the other part will contain the program which will run in the normal way – the background. In order to divide the memory in this way a linker (or loader) program which provides control over the locations in which the code is located and which allows several segments of code to be combined is required. A suitable linker is the Microsoft L80, used in conjunction with the M80 assembler and FORTRAN 80.

To make use of this approach the tasks to be performed must be divided into two categories: time critical and non-time critical. The time-critical tasks are run inside the interrupt routine as the foreground and the other tasks are run in the background. Considering the system described in Chapter 5, the time-critical tasks are: the acquisition of the plant data, output of plant control and calculation of the control outputs. These would have to be gathered together as the foreground task: the display to the operator and the input from the operator become the single background task. It is possible to write the bulk of both tasks in a high-level language as is shown in the skeleton programs given below.

* * *

Example 6.2 Simple foreground/background system

Foreground task

```
;
; Interrupt routine forming main body of foreground task
;
      PUBLIC IRH0       ;make address of interrupt routine available
                        ;to external programs
;
;
      EXT  SR,RR,CON,TIME  ;external routines, addresses will be
                           ;provided by the linker
;
```

```
IRHO:      CALL SR     ;save all registers
           CALL CON    ;control routines (could be FORTRAN routines)
           CALL TIME   ;clock routine
           CALL RR     ;restore all registers
           EI
           RETI
           END
```

Background task

```
C
C  Body of backgound task
C
C Initialize routines for setting up the system
C
           CALL COMVAR(p1,p2, ....pn)
C
C Above is a routine which transfers the addresses of variables
C p1,p2,..pn to the foreground program, it is only required if
C the control routines are written in assembler
C
           SECS=0
           DISPTM=0
C
C Set up interrupts and clock interval
C
      CALL INTP(IBASE,ICHAN,ICT)
C
C Main body of background
C
100        CALL TIME(SECS)
           IF SECS>DISPTM THEN 200
           CALL KEYB(STATUS)
           IF STATUS THEN 300
           GOTO 100
C
200        CALL DISPLAY
           CALL TIME(SECS)
           DISPTM=SECS+5
           GOTO 100
C
300        CALL KYBDIN
           GOTO 100
C
           END
```

*   *   *

In the above program example it is assumed that the sub-routines CON, SR, RR, INTP, COMVAR, and TIME are located in the foreground partition as is shown in Figure 6.6 and that their location has been determined by the programmer. In fact, provided that the routine used to set up the interrupts – INTP – finds within itself space for the interrupt response table, loads the address of this table into the

Simple Foreground/Background Operating System

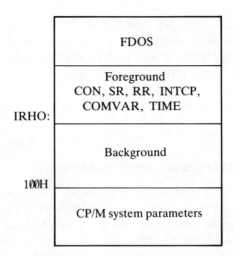

Fig. 6.6 Example of foreground/background partition.

Z80 I register, into the CTC chip, and also places the address of IRH∅: in the table, then the allocation of memory can be left to the linker. If allocation is left to the linker then there will be no identifiable contiguous foreground or background memory areas.

It is also assumed in the above that the controller parameters are held in COMMON and hence are available to both the keyboard input sub-routine (KYBDIN) and the control routine (CON). If the control routine is written in assembler in order to get faster execution, the background program is allowed to determine the location of the storage area for the parameters and the address of the actual locations is transferred to the foreground control sub-routine using COMVAR.

6.3.1 General foreground/background monitors

A wide range of foreground/background monitors have been developed; perhaps the best known is the DEC RT/11 operating system. Some provide little more than the CP/M modification described above, i.e., they allow interrupt routines to be used and guarantee that critical operations within the operating system will be protected from disturbance caused by the use of interrupts. Others provide extensive facilities, including the ability to have multiple tasks in both the background and foreground with the ability to roll tasks in and out of memory. This type are perhaps better described as multi-tasking operating systems with only two priority levels. Many systems provide some form of memory protection barrier between the foreground and background systems, either through software or hardware mechanisms. The provision of memory protection is particularly desirable in systems which allow program development and debugging to be a background activity.

Foreground/background systems often restrict the use of particular devices either to the foreground or to the background, but not to both, i.e., a printer may be allocated either to the foreground or to the background, but not to both: if a printer is required for both, two printers must be provided. Where such restrictions are not imposed, the problem of IO device-sharing becomes similar to that which occurs with multi-tasking operating systems and is discussed below.

6.4 REAL-TIME MULTI-TASKING OPERATING SYSTEMS

As was discussed in Chapter 5, the natural way to structure a typical computer control system is in the form of a number of different tasks which all apparently run in parallel. The implementation of a design based on this approach is made easier if an operating system which supports multi-tasking can be used. The traditional real-time operating systems are based on the assumption that all the tasks in the system will be executed on a single CPU or processor. In this section the same assumption will be made.

Confusion can arise between multi-user or multi-programming operating systems and multi-tasking operating systems. The function of a multi-user operating system is illustrated in Figure 6.7: the operating system ensures that each user can run

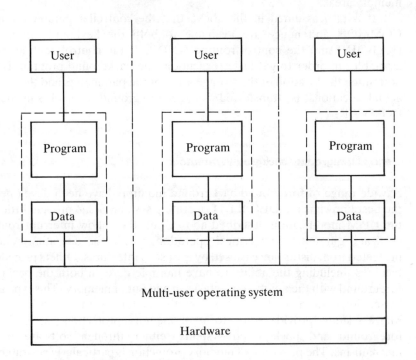

Fig. 6.7 Multi-user operating system.

a single program as if they had the whole of the computer system for their program. Although at any given instance it is not possible to predict which user will have the use of the CPU or even if the user's code is in the memory, the operating system ensures that one user program cannot interfere with the operation of another user program. Each user program runs in its own protected environment; a primary concern of the operating system is to prevent one program corrupting another, either deliberately or through error. In a multi-tasking operating system it is assumed that there is a single user and that the various tasks are to co-operate to serve the requirements of the user. Co-operation will require that the tasks communicate with each other and share common data. This is illustrated in Figure 6.8. In a good multi-tasking operating system the way in which tasks communicate and share data will be regulated such that the operating system is able to prevent inadvertent communication or data access (arising through an error in the coding of one task) and hence protect data which is private to a task. (Note that deliberate interference cannot be prevented – the tasks are assumed to be co-operating.)

A fundamental requirement of an operating system is to allocate the resources of the computer to the various activities which have to be performed; in a real-time

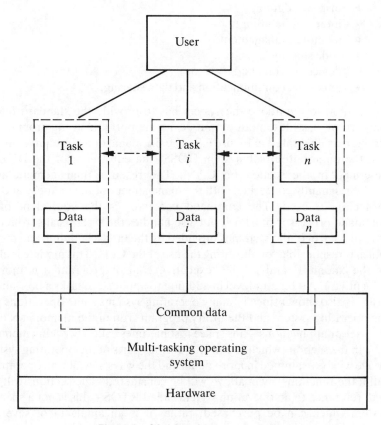

Fig. 6.8 Multi-tasking operating system.

operating system this allocation procedure is complicated by the fact that some of the activities are time-critical and hence have a higher priority than others. There must therefore be some means of allocating priorities to tasks and of scheduling allocation of CPU time to the tasks according to some priority scheme.

A task may use another task, i.e., it may require certain activities which are contained in another task to be performed and it may itself be used by another task. Thus tasks may need to communicate with each other. The operating system therefore has to have some means of enabling tasks, either to share memory for the exchange of data or to provide a mechanism by which tasks can send messages to each other. In addition, tasks may need to be invoked by external events; hence the operating system must support the use of interrrupts.

Similarly, tasks may need to share data and they may require access to various hardware and software components; hence there has to be a mechanism for preventing two tasks from attempting to use the same resource at the same time.

In summary, a real-time multi-tasking operating system (RTMTOS) has to support the resource sharing and the timing requirements of the tasks and the functions can be divided as follows:

- task scheduling;
- interrupt handling;
- memory management;
- code sharing;
- device sharing; and
- inter-task communication and data sharing.

In addition to the above the system has to provide the standard features such as support for disk files, input/output device support and utility programs. The typical structure is illustrated in Figure 6.9. The file manager is the equivalent of the BDOS and the input/output sub-system (IOSS) the equivalent of the BIOS which were described in the section on CP/M. The resource management and allocation module is mainly concerned with the management of the memory and with memory protection if used. The command processor is the equivalent of the console command processor of CP/M, but also handles the system calls from the tasks. The overall control of the system is provided by the task scheduling and dispatch module which is responsible for allocating the use of the CPU. This module is also referred to as the 'monitor' and as the 'executive control program' (or more simply the 'executive'). At the user level in addition to application tasks a box labelled 'systems tasks' is also shown since in many operating systems some operations performed by the operating system and the utility programs run in the memory space allocated to the user or applications – this space is sometimes called 'working memory'.

In the sections which follow, the various parts of the operating system described above will be examined in more detail and the ways in which an operating system can fulfill the functions given above will be considered. The sections will be illustrated with reference to an operating system called RTOS – this is not a specific operating system which can be purchased; rather it is an amalgam of several traditional real-time operating systems.

Task Management

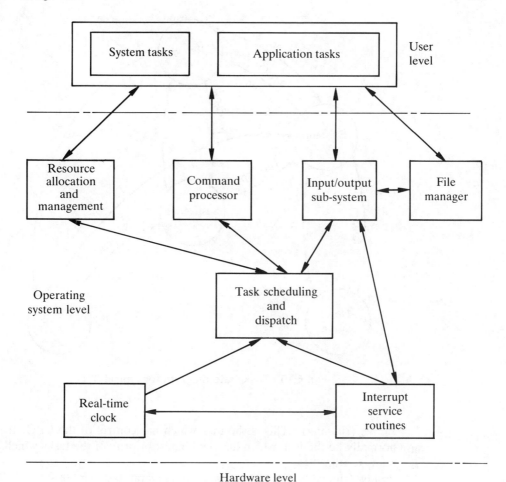

Fig. 6.9 General outline of real-time operating system.

6.5 TASK MANAGEMENT

6.5.1 Task states

On a computer system with only one processor only one task can run at any given time, hence the other tasks must be in some state other than running. The number of other states, the names given to those states and the transition paths between the different states vary with the operating system. A typical state diagram is given in Figure 6.10 and the various states are described below (names in parentheses are commonly used alternatives).

196   *Operating Systems*

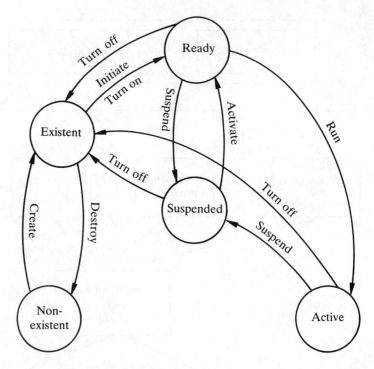

Fig. 6.10   Task state diagram for typical RTOS.

1. Active (running)   This is the task which has control of the CPU. It will normally be the task which has the highest priority of the tasks which are ready to run.
2. Ready (runnable, on)   There may be several tasks in this state. The attributes of the task and the resources required to run the task must be available for the task to be placed in the 'ready' state.
3. Suspended (waiting, locked out, delayed)   The execution of tasks placed in this state has been suspended because the task requires some resource which is not available or because the task is waiting for some signal from the plant, e.g., input from the analog-to-digital converter, or the task is waiting for the elapse of time.
4. Existent (dormant, off)   The operating system is aware of the existence of this task, but the task has not been allocated a priority and has not been made runnable.
5. Non-existent (terminated)   The operating system has not as yet been made aware of the existence of this task, although it may be resident in the memory of the computer.

The status of the various tasks may be changed by actions within the operating system – a resource becoming available or unavailable – or by commands from the

Table 6.6 RTOS task state transition commands

OFFC01	Turn off the task leaving the memory marked as occupied.
OFFC02	Turn off the task leaving the memory marked as unoccupied.
DELC01	Delay the task leaving the memory as occupied; delay is calculated using current value of time.
DELC02	Delay the task leaving the memory as unoccupied; delay is calculated as for DELC01.
DELC03	Delay the task leaving the memory occupied; the delay is calculated by adding the delay to the value of time stored in the task descriptor.
TPNC01	Turn the task on; will be accepted if the task is ON, OFF or DELAYED; either the ON constant can be placed in the task descriptor or a specified turn-on time.
TPNC02	Turn on the task; will only be accepted if the task is in the OFF state.
TPNC03	Run the task immediately regardless of priority; will be accepted if the task is ON, OFF or DELAYED.

application tasks. A typical command is TURN ON(ID) – transfer a task from 'existent' to 'ready' state where ID is the name by which the task is known to the operating system. A typical set of commands is given in Table 6.6. It should be noted that the transition from 'ready' to 'active' can only be made at the behest of the dispatcher.

6.5.2 Task descriptor

It should be evident from the discussion of task states and then from consideration of the interrupt mechanisms in Chapter 3 that the operating system must retain some status information about each of the tasks in the system. The information is held in a task descriptor (TD) – also referred to as a process descriptor (PD) or task control block (TCB) or task data block (TDB). It consists of an area of memory associated with each task. The information held in the TD will vary from system to system, but will typically consist of the following:

- task identification (ID);
- task priority (P);
- current state of task; and
- area to store volatile environment.

If the system permits tasks to be transferred between fast access memory and backing store (disk), information on the location of the task image on the disk and the area of fast memory occupied will be required. This could take the form of a file control block as described for the CP/M system. The TD will need to record whether or not the task is in fast access memory.

The TD for a task can be held in several ways; the two most common are tables and lists. The table method allocates fixed areas of memory for the various tables and the information for a given task is stored in the tables. The task descriptor is made up from the information drawn from the tables. This is shown in Figure 6.11 for RTOS operating system and the associated tables are shown in Figure 6.12.

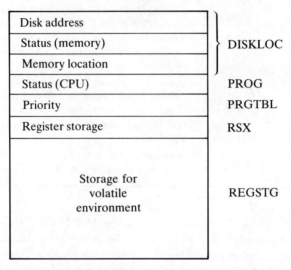

Fig. 6.11   RTOS task descriptor.

The information held in the DISKLOC table contains the address of the task on the disk, the location in memory when it is loaded and memory status information. This includes such information as whether it can be temporarily swapped to disk, or should remain permanently in memory once loaded; if it is currently in memory, whether the data has to be saved if the task is swapped out of memory, etc.

The PROG table holds information on the current status of the task: i.e., ON, READY, DELAYED, LOCKED OUT, OFF, etc. This particular operating system distinguishes between tasks which are suspended awaiting completion of some operating system request such as an output to a device – these tasks are marked as locked out – and tasks which are suspended until a particular time, which are marked as delayed. The task state diagram for the system is shown in Figure 6.8. The PRGTBL is used by the dispatcher to determine the order in which to search for the next task to run, hence the order of the entry of the task's number (task ID) in the table determines the priority of the task. The task which is placed in the first location of the table is the highest priority task. This provides a flexible system of allocating priorities and enables the priority of a task to be changed by changing the order of the entries in the table.

There is a problem with a table structure in that to move a task from a low priority to a high priority requires the re-ordering of the whole table. One simple way to avoid this is to allocate a dummy task at a high priority position and simply exchange the dummy task with the low priority task. In practice, however, frequent

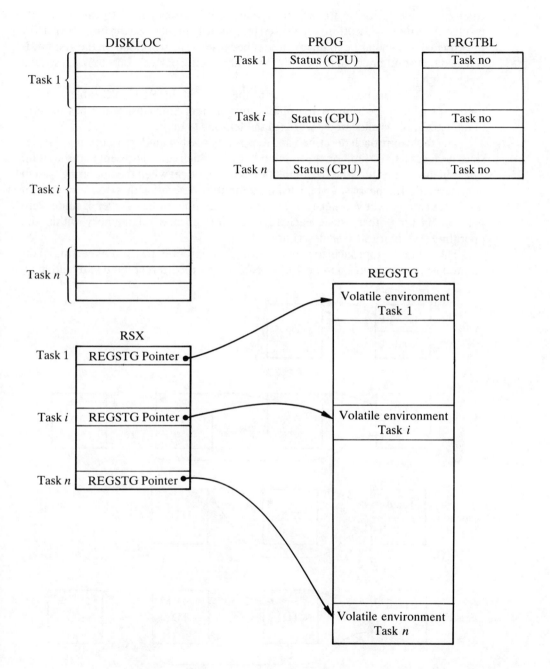

Fig. 6.12 RTOS tables for storage of task descriptors.

changes in task priority are not recommended: they can lead to unpredictable results. Priority re-allocation can be used temporarily to increase the priority of a low priority task which has been unable to run because of the workload on the system. It can also be useful under alarm conditions when it may be necessary to run a special task at high priority.

The RSX table contains a pointer to the entry in the REGSTG table which is used to provide space to store the volatile environment. Each task which has an entry in the RSX table is allocated 8 words in the REGSTG table.

The table approach described above is rarely used in modern operating systems since it places a fixed limit on the number of tasks which can be present in the system. The limit arises because the size of the tables has to be set when the operating system is generated. In the past, when using computer systems with small amounts of fast-access memory it was necessary to limit the number of tasks, even if some were held on backing store, since each task requires an area of memory within the operating system for its task descriptor.

The modern approach to holding task descriptors is to use linked lists. This method is illustrated in Figure 6.13: for each task a record is created and held in an

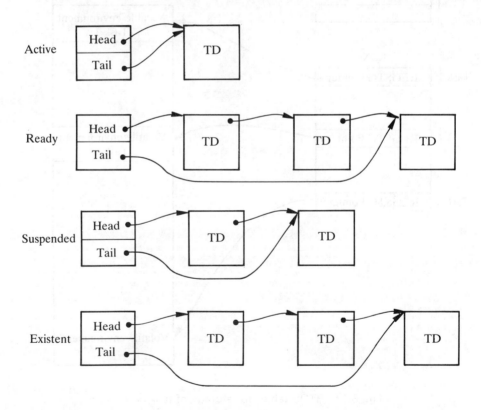

Fig. 6.13 List structure for task states.

area of memory; one entry in the record, usually the first, is used as a pointer to the next entry in the list, i.e., it contains the memory address of the next record. The operating system keeps one central table which has entries giving the pointers to the heads and tails of the various lists – runnable tasks, suspended tasks, etc. To move a task from one list to another, changes are made to the list pointers. In a real-time system the tasks would probably be held in priority order in the list so that, for example, the dispatcher would simply have to check the first element, i.e., the task descriptor of the first task in the list, to see if it was of higher priority than the currently active task.

The advantage of the list structure is that the actual task descriptor can be located anywhere in the memory and hence the operating system is not restricted to a fixed number of tasks. It is possible with this type of operating system to allow the dynamic creation and destruction of tasks. However, for a fixed number of high priority and frequently run tasks the procedure involved in searching for work can take longer with a list structure than it does with a table structure.

6.6 TASK DISPATCH AND SCHEDULING

The primary function of the segment of the operating system which deals with task dispatch and scheduling is the allocation of CPU time to the various tasks. The scheduler deals with the high-level allocation decisions: it receives information on the availability of the various resources of the system, e.g., printers, disk drives, etc., and it then moves tasks from the suspended state to the ready state. The scheduler may also keep track of low priority tasks which have been waiting for a long time and change their priority to ensure that they run.

At a lower level than the scheduler is the dispatcher (sometimes called the low-level scheduler) which swaps the tasks by saving the status of the current task and establishing the next task to run. It is also the responsibility of the dispatcher to make the short-term decisions in response to, e.g., interrupts from an input/output device or from the real-time clock.

6.6.1 Priority levels

In a real-time system the designer has to assign priorities to the tasks in the system. The priority will depend on how quickly a task will have to respond to a particular event. An event may be some activity of the process or may be the elapsing of a specified amount of time. It is usual to divide the tasks into three broad levels of priority as shown in Figure 6.14.

1. Interrupt level At this level are the service routines for the tasks and devices which require very fast response – measured in milliseconds. One of these tasks will be the real-time clock task and clock level dispatcher.

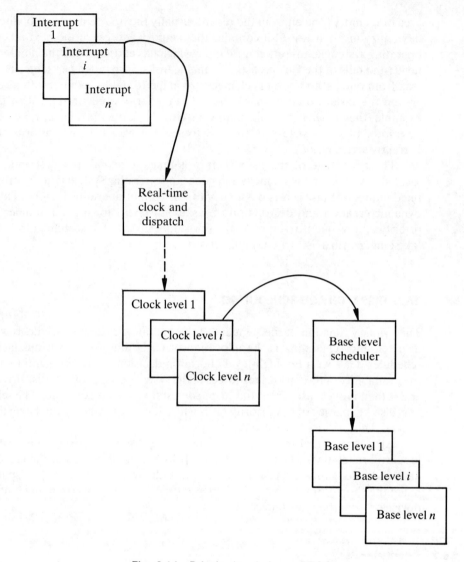

Fig. 6.14 Priority levels in an RTOS.

2. **Clock level** At this level are the tasks which require repetitive processing, such as the sampling and control tasks, and tasks which require accurate timing. The lowest priority task at this level is the base level scheduler.
3. **Base level** Tasks at this level are of low priority and either have no deadlines to meet or are allowed a wide margin of error in their timing. Tasks at this level may be allocated priorities or may all run at a single priority level – that of the base level scheduler.

Task Dispatch and Scheduling 203

The existence of interrupt-level and clock-level tasks implies the use of a preemptive scheduling strategy, by which is meant than when a task wishes to become active, and it has a priority higher than the currently running task, then the currently running task will be stopped and the higher priority task will be run. A purely non-preemptive strategy would imply that all tasks at any level would run to completion without interruption and only when a task completed would a rescheduling of tasks take place.

Few operating systems run using a totally non-preemptive scheduling scheme. Many non-real-time systems allow interrupts from device handlers and from a clock and reschedule at specified clock intervals. They also reschedule when a task makes a request for some system resource, e.g., a printer.

* * *

Example 6.3   Preemptive and non-preemptive scheduling

A system receives an alarm signal interrupt from a process and the action which is required in response to the alarm is to run an alarm alert task which is a high priority base-level task. In response to the alarm interrupt, the interrupt service routine for the alarm will, by some mechanism, cause the alarm alert task to be placed in the runnable queue; there are then two actions which it can take, (a) return to the interrupted task, or (b) jump to the dispatcher. If a return to the interrupted task is made then the alarm alert task will not be run until the system reschedules either at the system rescheduling interval or because the running task terminates or becomes suspended waiting for a system resource. This would be termed a non-preemptive strategy. However, if a jump is made directly to the dispatcher from the interrupt service routine, then if the alarm alert task is of higher priority than the interrupted task it will be run immediately and the interrupted task will have been preempted.

6.6.2   Interrupt level

As we have already seen an interrupt forces a rescheduling of the work of the CPU and the system has no control over the timing of the rescheduling. Because an interrupt-generated rescheduling is outside the control of the system it is necessary to keep the amount of processing to be done by the interrupt service routine to a minimum. It is usual for the interrupt service routine to do sufficient processing to preserve the necessary information and to pass this information to a further service routine which operates at a lower priority level, either clock level or base level. Interrupt service routines have to provide a mechanism for task swapping, i.e., they have to save the volatile environment. On completion the routine will either simply restore the volatile environment and hence will return to the interrupted task, or it may exit to the dispatcher.

Within the interrupt level of tasks there will be different priorities and there will have to be provision for preventing interrupts of lower priority interrupting higher priority interrupt tasks. On most modern computer systems there will be hardware to assist in this operation (see Chapter 3).

6.6.3 Clock level

One interrupt level task will be the real-time clock service routine which will be entered at some interval, usually determined by the required activation rate for the most frequently required task. Typical values are 20 to 200 ms. Each clock interrupt is known as a 'tick' and represents the smallest time interval known to the system. The function of the clock interrupt service routine is to update the time-of-day clock in the system – this may be simply to increment a counter in a memory location – and to transfer control to the dispatcher. The dispatcher selects which task is to run at a particular clock tick.

Clock level tasks divide into two categories:

1. Cyclic   These are tasks which require accurate synchronization with the outside world.
2. Delay   These tasks simply wish to have a fixed delay between successive repetitions or to delay their activities for a given period of time.

6.6.4 Cyclic tasks

The cyclic tasks are ordered in a priority which reflects the accuracy of timing required for the task with those which require high accuracy being given the highest priority. Tasks of lower priority within the clock level will have some jitter since they will have to await completion of the higher level tasks.

* * *

Example 6.4   Cyclic tasks

Three tasks A, B and C are required to run at 20 ms, 40 ms and 80 ms intervals (corresponding to 1 tick, 2 ticks and 4 ticks, if the clock interrupt rate is set at 20 ms). If the task priority order is set as A, B and C with A as the highest priority then the processing will proceed as shown in Figure 6.15a with the result that the tasks will be run at constant intervals. It should be noted that using a single CPU it is not possible to have all the tasks starting in synchronism with the clock tick. All but one of the tasks will be delayed relative to the clock tick, but the interval between successive invocations of the task will be constant.

If the priority order is now rearranged so that it is C, A and B then the activation diagram is as shown in Figure 6.15b and every fourth tick of the clock there will be a delay in the timing of tasks A and B. In practice there is unlikely to be any justification for choosing a priority order C, A and B rather than A, B and C. Usually the task with the highest repetition rate will have the most stringent timing requirements and hence will be assigned the highest priority.

A further problem which can arise is that a clock-level task may require a longer time than the interval between clock interrupts to complete its processing. (Note that for overall satisfactory operation of the system such a task cannot run at a high repetition rate.)

* * *

Task Dispatch and Scheduling

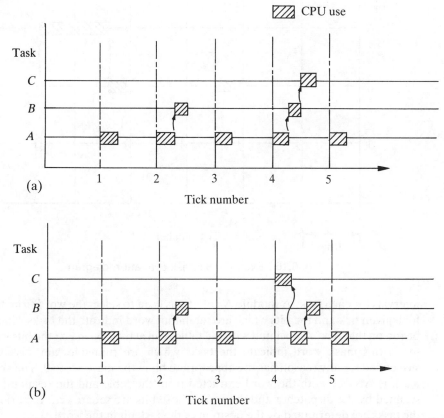

Fig. 6.15 Example 6.2 task activation diagram. (a) Priority A,B,C; (b) priority C,A,B.

Example 6.5 Timing of cyclic tasks

Assume that in Example 6.1 task C takes 25 ms to complete, task A takes 1 ms and task B takes 6 ms. If task C is allowed to run to completion then the activity diagram will be as shown in Figure 6.16 and task A will be delayed by 11 ms at every fourth invocation. It is normal therefore to divide the cyclic tasks into high priority tasks, which are guaranteed to complete within the clock interval, and lower priority tasks which can be interrupted by the next clock tick.

Careful design of the real-time clock handler, which acts as the dispatcher for the system and controls the activation of the clock-level tasks, is required since it is run at frequent intervals. Particular attention has to be paid to the method of selecting the tasks to be run at each clock interval. If a check of all tasks were to be carried out, the overheads involved could become significant.

One method which is used is illustrated in Figure 6.17. Each task is allocated to one bit in a word and a table is set up with a number of entries (words) corresponding to the number of ticks which will be required to produce the invocation of all the tasks. This number, N, corresponds to the lowest common multiplier of the timing

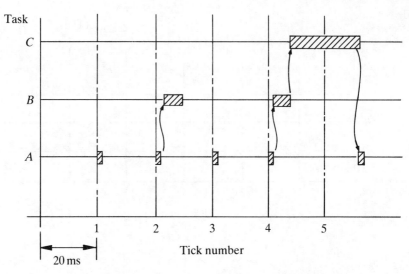

Fig. 6.16 Example 6.5 task activation diagram.

intervals for each task. A modulo N counter is used to select the word from the table for a given tick and the bits which are set in that word indicate the tasks which are to be run on that tick. Associated with the table is an activation mask: the bits which are set in this mask word indicate the tasks which are runnable; any task which is suspended for any reason has the appropriate bit in the mask set to 0. The activation mask is ANDed with the word extracted from the table and the resultant word is scanned by the dispatcher and the tasks whose bits are set are run: the priorities of the tasks are determined by the position of the task bit in the word.

* * *

Example 6.6 Control table for cyclic tasks

It is assumed that there are seven clock-level tasks with repetition intervals of 1, 2, 4, 5, 10 and 20 ticks. The lowest common multiplier is therefore 20 and, as is shown in Figure 6.17, a table with 20 entries is required. It is assumed that all the tasks will be required to run at tick 0 so all the task bits are set. The modulo 20 counter is incremented at each clock tick and its value is used as an index to the table entries; at tick 4 it points to entry four and hence tasks A, B and C are due to run. However, the activation mask has the bit corresponding to task C reset, hence for some reason C is not ready to run so only tasks A and B will be run.

* * *

The advantage of this method is that it is fast in that the counter points directly to the word containing the tasks to be run at the particular clock interval. The disadvantages are that a large table will be required if the cycle times for the tasks are such that the lowest common multiplier is a large number. If in Example 6.6 the

Task Dispatch and Scheduling

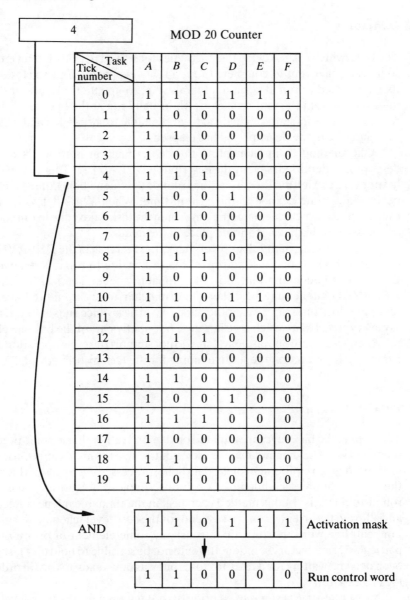

Fig. 6.17 Clock-level dispatching technique.

repetition frequency of task C had been set at 3 ticks rather than 4, the lowest common multiplier would have been 60 not 20 and hence a table with 60 entries would have been required. The method also becomes complicated if the number of clock level tasks exceeds the word length of the system.

6.6.5 Delay tasks

The tasks which wish to delay their activities for a fixed period of time, either to allow some external event to complete (e.g., a relay may take 20 ms to close) or because they only need to run at certain intervals (e.g. to update the operator display), use information from the real-time clock but usually run at the base level. When a task requests a delay its status is changed from runnable to suspended and it has to remain suspended until the delay period has elapsed.

One method of implementing the delay function is to use a queue of task descriptors, identified, say, by the description DELAYED. This queue is an ordered list of task descriptors: the task at the front of the queue being that which contains a next running time nearest to the current time. When a task delays itself it calls an executive task which calculates the time when the task is next due to run and inserts the task descriptor in the appropriate place in the queue.

A task running at the clock level checks the first task in the DELAYED queue to see if it is time for that task to run. If the task is due to run it is removed from the DELAYED queue and placed in the runnable queue. The task which checks the DELAYED queue may be either the dispatcher which is entered every time the real-time clock interrupts or another clock-level task which runs at frequent intervals, say every 10 ticks, in which case it is then frequently part of the base-level scheduler.

Many real-time operating systems do not support the cycle operation and the user has to create an accurate repetitive timing for the task by using the delay function.

6.6.6 Base level

The tasks at the base level are initiated on demand rather that at some predetermined time interval. The demand may be user input from a terminal, some process event or some particular requirement of the data being processed. The way in which the tasks at the base level are scheduled can vary; one simple way is to use time slicing on a round-robin basis. In this method each task in the runnable queue is selected in turn and allowed to run either until it suspends or the base level scheduler is again entered. For real-time work in which there is usually some element of priority this is not a particularly satisfactory solution. It would not be sensible to hold up a task, which had been delayed waiting for a relay to close, but was now ready to run, in order to let the logging task run.

Most real-time systems use a priority strategy even for the base-level tasks. This may be either a fixed level of priority or a variable level. The difficulty with a fixed level of priority lies in determining the correct priorities for satisfactory operation. The ability to change priorities dynamically allows the system to adapt to particular circumstances. Dynamic allocation of priorities can be carried out using a high-level scheduler or can be done on an ad hoc basis from within specific tasks.

The high-level scheduler is an operating system task which is able to examine the use of the system resources; it may, for example, check how long tasks have been waiting and increase the priority of the tasks which have been waiting a long time. The

Task Dispatch and Scheduling

difficulty with a high-level scheduler is that the algorithms used can become complicated and hence the overhead in running can become significant.

An alternative is to adjust priorities in response to particular events or under the control of the operator. For example, alarm tasks will usually have a high priority and, during an alarm condition, tasks such as the log of plant data may be delayed with the consequence that the output of the log lags behind real-time. (Note the data will be stored in buffer areas inside the computer.) In order that the log can catch up with real-time quickly it may be advisable to increase, temporarily, the priority of the printer output task.

6.6.7 System commands which change task status

The range of system commands affecting task status provided in a real-time operating system varies with the operating system. The task states and task commands for the RTOS operating system are shown as an example in Figure 6.18. Fuller details of the commands are given in Table 6.6. It should be noted that this system distinguishes between tasks which are suspended waiting the passage of time – these tasks are marked as delayed – and those tasks which are waiting for an event or a system resource – these are marked as locked out.

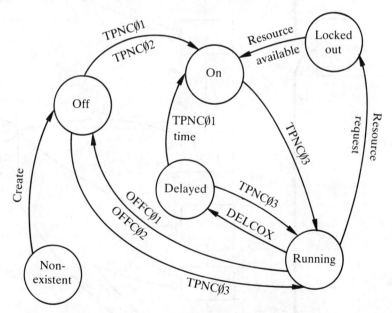

Fig. 6.18 RTOS task state diagram.

The system does not support explicitly base level tasks; however the lowest four priority levels of the clock level tasks can be used to create a base level system. A so-called free time executive (FTX) is provided which if used runs at priority level

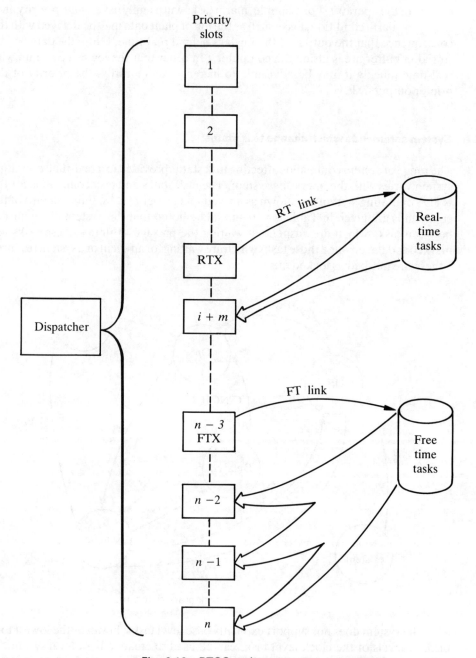

Fig. 6.19 RTOS task structure.

$n - 3$ (see Figure 6.19) where n is the lowest priority task number. The FTX is used to run tasks at priority levels $n - 2$, $n - 1$ and n; it provides support for the chaining of tasks. The dispatcher is unaware of the fact that tasks at these three priority levels are being changed: it simply treats whichever tasks are in the lowest three priority levels as low priority tasks. Tasks run under the FTX do not have access to the system commands (except OFFC∅1, i.e., turn task off).

When the RTOS is generated a decision has to be made as to the number of task priority levels which will be permitted (maximum number is 256), but to enable a much larger number of tasks to be run system commands which permit the chaining of real-time (as opposed to free-time) tasks are provided. These commands permit the creation, at any priority level, of a simple real-time executive (RTX) which can be used to form a number of base level systems. As with the FTX the additional tasks are resident on the disk – in the free-time system the tasks have names; in the RTX system they are given numbers – and the system commands are used to inform the clock-level dispatcher of their existence. The command takes the form:

```
CALL RTLINK
LDX  N           ; priority level number
LDX  M           ; task number
LDX  I           ; variable used to return error code
```

The priority number is the priority level at which the task is to be run; it must be a priority level lower than that of the task which makes the call. M is the number of the task and I is a variable which will contain an error code if the RTLINK call fails. Using the system a task running at, say, priority level 30 could run a whole series of tasks at say priority level 40. It should be noted that once a task is installed in a priority slot it cannot be removed until the task itself releases the slot by means of a system command RTRELE. Tasks running under RTX can access any of the system commands.

6.6.8 Dispatcher: search for work

The dispatcher/scheduler has two entry conditions:

1. The real-time clock interrupt and any interrupt which signals the *completion* of an input/output request;
2. A task suspension due to a task delaying, completing or requesting an input/output transfer.

In response to the first condition the scheduler searches for work starting with the highest priority task and checking each task in priority order (see Figure 6.20). Thus if tasks with a high repetition rate are given a high priority they will be treated as if they were clock-level tasks, i.e., they will be run first during each system clock period. In response to the second condition a search for work is started at the task with the next lowest priority to the task which has just been running. There cannot be another higher priority task ready to run since a higher priority task becoming ready always preempts a lower priority running task.

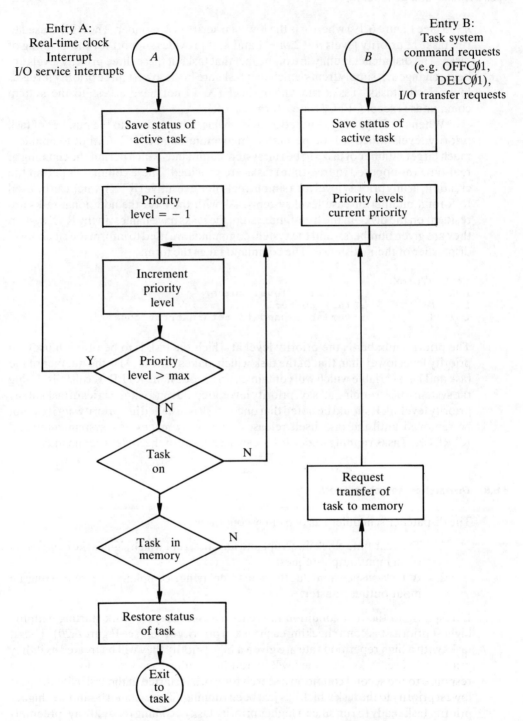

Fig. 6.20 RTOS search for work by the dispatcher.

Task Dispatch and Scheduling 213

The system commands for task management are issued as calls from the assembly level language and the parameters are passed either in the CPU registers or as a control word immediately following the call statement.

* * *

Example 6.7 Use of RTOS system calls

As an example, the system whose outline design is given in Figure 5.14 will be used. It is assumed that the control, display and operator input programs are to be run as separate tasks with priorities 1, 10, 20 respectively. The control task has to run at 40 ms intervals and the display update task at 5 s intervals. The system clock is set at 20 ms and hence the control task has to run every 2 system ticks. The outline structure of the system is given below.

```
        TASK MAIN
;
; Starts up the system by creating the various tasks and setting
;      them to the ON condition
;
        CREATE(CONTROL,1,STCTRL)
        CREATE(DISPLAY,10,STDISP)
        CREATE(OPERATOR,20,STOPR)
;
; STCTRL, STDISP, STOPR are common sysmbols which define the
; starting locations for each task, the values will be
; inserted by the linker/loader.
;
        LDA TIME        ;TIME is system variable which gives current
                        ; time
        CALL TPNC01
        FCB  1          ; turn on control task
;
        LDA 0
        CALL TPNC02
        FCB  10         ; turn on display task
;
        LDA 0
        CALL TPNC02
        FCB  20         ; turn on operator input task
;
        CALL OFFC02     ; terminate main task
;
        END
```

In the above code by using TPNCØ1 to turn on the control task the current value of time is placed in the task descriptor and hence the task can be synchronized to the clock.

```
        TASK CONTROL
;
        .....
        main body of task
        ....
;
```

```
              CALL  DELC03
              FCB   0,2        ; set next time for running to previous time
                               ; plus two ticks
        ;
              END

              TASK  DISPLAY
        ;
              ....
              main body of task
              ....
        ;
              CALL  DELC02
              FCB   5,0        ; delay task by 5 seconds
        ;
              END
```

The difference between using DELC∅3 and DELC∅2 in the above task segments is that DELC∅3 adds the delay increment to the value of time stored in the task descriptor; this time is the time at which the task was last due to run. The use of DELC∅3 therefore provides a means of running tasks in a cyclic mode at clock level. The DELC∅2 command adds the delay value to the current time and stores the result in the task descriptor, hence the delay is calculated, not from when the task was last due to run, but from the time at which the delay command was issued.

In designing real-time systems it can be important to know how the scheduler of the operating system to be used searches for work. In the system described above, the scheduler searches in strict priority order and hence the overheads in terms of the time spent searching for work will be increased if some of the tasks which have a high priority rarely run. A careful assessment of task priority is required and particular attention will have to be paid to alarm action tasks. Such tasks are normally accorded high priority, but it is hoped that they will rarely be required. One solution with the above system which avoids having a group of high priority, but rarely run, alarm tasks is to make use of the TPNC∅3 command. The alarm action tasks can be given low priority, but can be made to run immediately if the alarm scanning routine uses the TPNC∅3 call to invoke the appropriate task.

6.6.9 Deadlock

The competition for resources among tasks can result in the condition known as *deadlock*. Suppose Task A has acquired exclusive use of Resource X and now requests Resource Y, but between A acquiring X and requesting Y, Task B has obtained exclusive use of Y and has requested use of X. Neither task can proceed, since A is holding X and waiting for Y and B is holding Y and waiting for X. The system is said to be deadlocked. The detection of deadlock or the provision of resource-sharing commands in such a way as to avoid deadlock is the responsibility of the operating system [see Lister p. 94–7 for discussion of deadlock avoidance and detection mechanisms].

6.7 MEMORY MANAGEMENT

Since the majority of control application software is static – the software is not dynamically created or eliminated at run time – the problem of memory management is simpler than for multi-programming, online systems. Indeed as the cost of computer hardware, both processors and memory, reduces, many control applications use programs which are permanently resident in fast access memory.

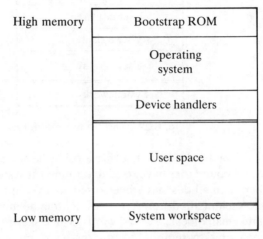

Fig. 6.21 Non-partitioned memory.

In systems where the control application software is permanently resident the memory can be divided as shown in Figure 6.21 in which the user space is treated as one unit and the software is linked and loaded as a single program into the user area. The information about the various tasks is conveyed to the operating system by means of a create task statement. Such a statement may be of the form

CREATE(TASKID,PRIORITY,STARTADDRESS)

this is the equivalent of the FORM statement in MASCOT, and it would result in the operating system creating a TD for the specified task.

An alternative arrangement is shown in Figure 6.22 where the available memory is divided into predetermined segments and the tasks are loaded individually into the various segments. The load operation would normally be carried out using an equivalent to the console command processor of CP/M. Usually with this type of system the entries in the TD (or the operating system tables) have to be made from the console using a memory examine and change facility.

The use of divided (partitioned) memory was widely used in many early real-time operating systems and it was frequently extended to allow several tasks to share one partition; the tasks were kept on the backing store and loaded into the appropriate partition when required. There was of course a need to keep any tasks in which timing was crucial, i.e., the type 2 tasks, in fast access memory permanently, but other tasks could be swapped between fast memory and backing store. The

```
┌─────────────────────┐
│    Bootstrap ROM    │
├─────────────────────┤
│      Operating      │
│       system        │
├─────────────────────┤
│   Device handlers   │
├─────────────────────┤
│   User task area 1  │
├─────────────────────┤
│   User task area 2  │
├─────────────────────┤
│   User task area 3  │
├─────────────────────┤
│   User task area 4  │
├─────────────────────┤
│   System workspace  │
└─────────────────────┘
```

Fig. 6.22 Partitioned memory.

difficulty with this method is, of course, in choosing the best mix of partition sizes. The partition size and boundaries have to be determined at system generation.

A number of methods have been used to overcome the problem of fixed partitions: one such method, referred to as floating memory, divided the available memory up into small blocks, e.g., 64 words. The tasks are installed on the backing store and, when a task is required to run, the operating system examines a map of memory and finds a contiguous area of memory which will hold the task. The task is loaded into the memory and the memory blocks occupied by the tasks are marked as occupied in the memory map. With this technique the area occupied by a task is closely related to the actual size of the task and not to some predetermined fixed partition size. A task which for some reason becomes suspended or delayed will have the memory area it occupies marked in the memory map as occupied but available, hence if another task becomes ready then the suspended task can be returned to backing store and the ready task loaded into its area. In this type of system information must be held in the task descriptor to indicate if the task can be swapped, since, e.g., the control task of Example 6.7 which has to run every 40 ms would have to be held permanently in memory in order to guarantee the sampling rate. A problem which is generated by this system is fragmentation of the available memory. It is possible to get small areas of free memory spread about the memory address space, none of the individual areas being large enough to take a task, but the combined areas could if it were possible to bring them together. It is therefore necessary to have some form of 'garbage' collection to bring dispersed areas into contiguous blocks.

Other systems which permit dynamic allocation of memory allow the tasks themselves to initiate program segment transfers, either by chaining or by overlaying, and task swaps. In chaining the task is divided into several segments which run sequentially. On completion of one segment the next segment is loaded from memory into the area occupied by the previous segment, and any data required to be passed is either held on the disk or in a common area of memory.

Memory Management

Task swapping involves one task invoking another task; the first task is transferred to backing store and the second task brought into memory and made available to run. The procedure is shown in Figure 6.23. Task 1 invokes Task 5 by swapping it into Priority Level 41 and in turn Task 5 chains Task 6 into Level 41. Task 6 swaps Task 7 into Level 42. When Task 7 terminates the operating system returns control to Task 6 and when it terminates control is returned to Task 1. It should be

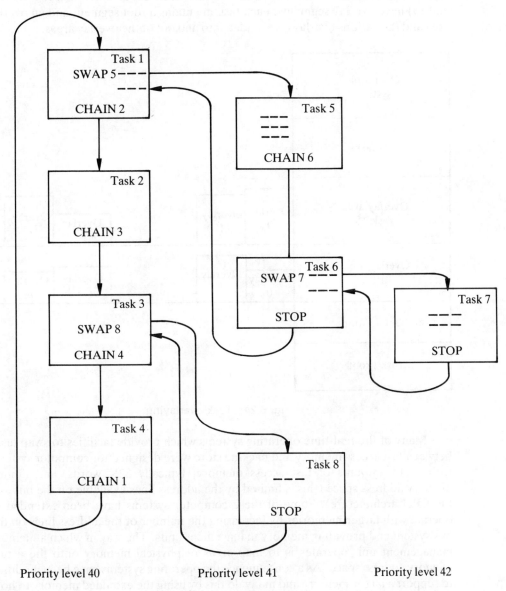

Fig. 6.23 Task chaining and swapping.

noted that Task 1 remains suspended until Task 6 terminates and, similarly, Task 6 is suspended until Task 7 terminates.

The difference between chaining and overlaying is that in overlaying a part of the task, the root task remains in memory and various segments are brought in sequence into an overlay area of memory. In a multi-tasking system there may be several different overlay areas each of which may be shared by several tasks. A typical arrangement is shown in Figure 6.24 in which it is assumed that two tasks (1 and 15) have overlay segments; each task maintains a root segment and an overlay area and the various overlays are loaded into and out of the overlay areas.

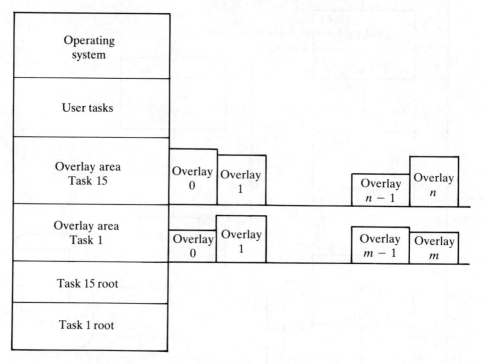

Fig. 6.24 Task overlaying.

Many of the real-time operating systems which provide facilities to swap tasks between fast access memory and backing store were designed for computer systems with small amounts of fast access memory, typically 32K words, and limited memory address space (this is limited by the address lines available on the bus and the CPU architecture). Some of these computer systems have been extended to operate with larger memories by increasing the number of the address lines on the bus system and providing memory management units. The way in which a memory management unit operates is to map areas of physical memory onto the actual memory address space. As a consequence the operating systems have been modified to support a larger memory and many do this by using the extended memory to hold the task overlays with a requirement that the root task still be located in the first segment of the memory.

6.8 CODE SHARING

In many real-time computer applications, the same actions will have to be carried out in several different tasks. In a conventional program the actions would be coded as a sub-routine and one copy of the sub-routine would be included in the program. In a multi-tasking system, either each task must have its own copy of the sub-routine or some mechanism must be provided to prevent one task interfering with the use of the code by another task. The problems which can arise can be illustrated as follows: Figure 6.25 shows two tasks which share the sub-routine A. If Task A is using the sub-routine but before it finishes some event occurs which causes a rescheduling of the tasks and Task B runs and uses the sub-routine, then when a return is made to Task A, although it will begin to use Subroutine S again at the correct place, the values of locally held data will have been changed and will reflect the information processed within the sub-routine by Task B.

If the sub-routine is large, or if there are many tasks which use the sub-routine, there can be a considerable saving of memory if only one copy of the sub-routine is present in the memory. Two methods can be used to enable this to be done:

- serially reusable code; and
- re-entrant code.

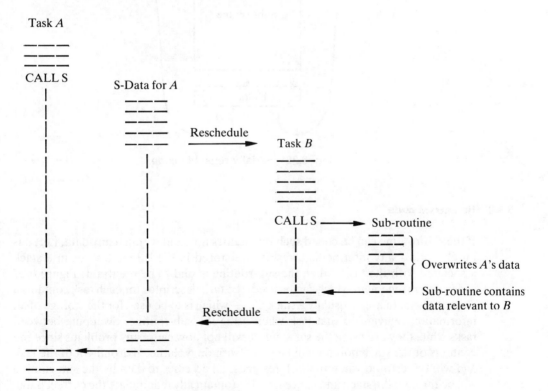

Fig. 6.25 Sharing a sub-routine in a multi-tasking system.

6.8.1 Serially reusable code

The sub-routine is written in such a way that the value of any local variable on entry to the routine has no effect on the actions of the routine. As is shown in Figure 6.26, some form of lock mechanism is placed at the beginning of the routine such that if any task is already using the routine, the calling task will not be allowed entry until the task which is using the routine unlocks it. The use of a lock mechanism to protect a sub-routine is an example of the need for mechanisms to support mutual exclusion when constructing an operating system. Mutual exclusion mechanisms are discussed in Chapter 7.

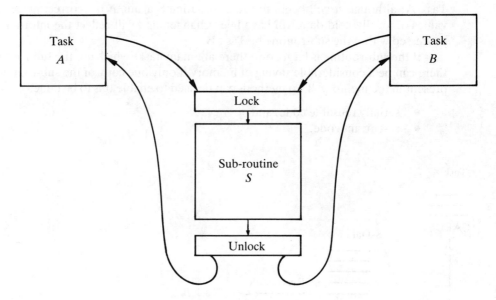

Fig. 6.26 Serially reusable code.

6.8.2 Re-entrant code

If the sub-routine can be coded such that it does not hold within it any data, i.e., it is purely code – any intermediate results are stored in the calling task or in a stack associated with the task – then the sub-routine is said to be reentrant. Figure 6.27 shows an arrangement which can be used: the task descriptor for each task contains a pointer to a data area – usually a stack area – which is to be used for the storage of all information relevant to that task when using the sub-routine. Swapping between tasks while they are using the sub-routine will not now cause any problems since the contents of the stack pointer will be saved with the volatile environment of the task and will be restored when the task resumes; all accesses to data by the sub-routine will be through the stack and hence it will automatically manipulate the correct data.

Code Sharing

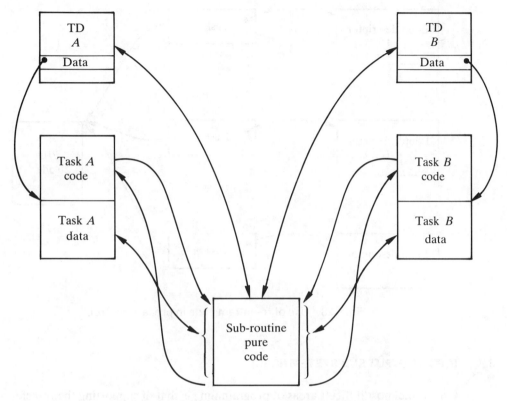

Fig. 6.27 Use of re-entrant code for code sharing.

Re-entrant routines can be shared between several tasks since they contain no data relevant to a particular task and hence can be stopped and restarted at a different point in the routine without any loss of information. The data held in the working registers of the CPU is stored in the relevant task descriptor when task swapping takes place.

Device drivers in conventional operating systems are frequently implemented using re-entrant code. Another application could be for the actual three-term (PID) control algorithm in a process control system with a large number of control loops. The mechanism is illustrated in Figure 6.28; associated with each control loop is a LOOP descriptor as well as a TASK descriptor. The LOOP descriptor contains information about the measuring and actuation devices for the particular loop, e.g., the scaling of the measuring instrument, the actuator limits, the physical addresses of the input and output devices, and the parameters for the PID controller. The PID controller code segment uses the information in the LOOP descriptor and the TASK to calculate the control value and to send it to the controller. The actual task is made up of the LOOP descriptor, the TASK segment and the PID control code segment. The addition of another loop to the system requires the provision of new loop descriptors; the actual PID control code remains unchanged.

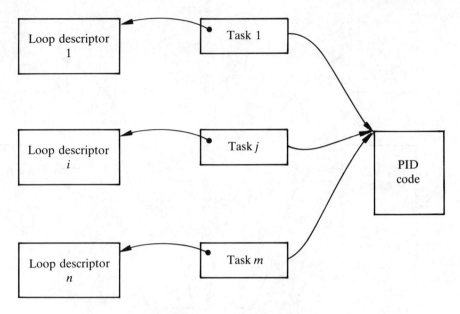

Fig. 6.28 Use of re-entrant code in process control.

6.9 INPUT/OUTPUT SUB-SYSTEM (IOSS)

One of the most difficult areas of programming is that of supporting the transfer of information to and from external devices. The availability of a well designed and implemented input/output sub-system (IOSS) in an operating system is essential for efficient programming. The presence of such a system enables the application programmer to perform input or output by means of system calls either from a high-level language or from the assembler. The IOSS handles all the details of the devices. In a multi-tasking system the IOSS should also deal with all the problems of several tasks attempting to access the same device.

A typical IOSS will be divided into two levels as shown in Figure 6.29. The I/O manager accepts the system calls from the user tasks and transfers the information contained in the calls to the device control block (DCB) for the particular device. The information supplied in the call by the user task will be, e.g., the location of a buffer area in which the data to be transferred is stored (output) or is to be stored (input); the amount of data to be transferred; the type of data, e.g. binary or ASCII; and the direction of transfer and the device to be used.

The actual transfer of the data between the user task and the device will be carried out by the device driver and this segment of code will make use of other information stored in the DCB. A separate device driver may be provided for each device or, as is shown in Figure 6.30, a single driver may be shared between several devices; however, each device will require its own DCB. The actual data transfer will usually be carried out under interrupt control.

Input/Output Sub-system (IOSS)

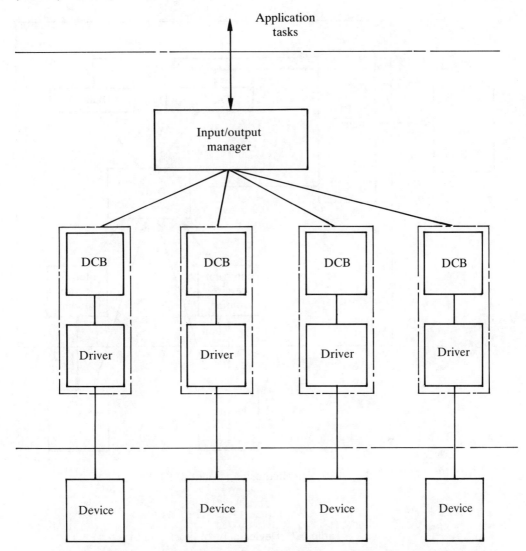

Fig. 6.29   General structure of IOSS.

Typically a DCB will contain the information shown in Table 6.7. The physical device name is the name by which the operating system recognizes the device and the type of device is usually given in the form of a code recognized by the operating system. The operating system will normally be supplied with DCBs for the more common devices. The DCBs may require modifying to reflect the addresses used in a particular system, although many suppliers adopt the policy of using standard addresses both for the physical address of the device on the bus and for the interrupt locations and interrupt service routines. The addition of non-standard devices will require the user to provide appropriate DCBs, but this task is usually made

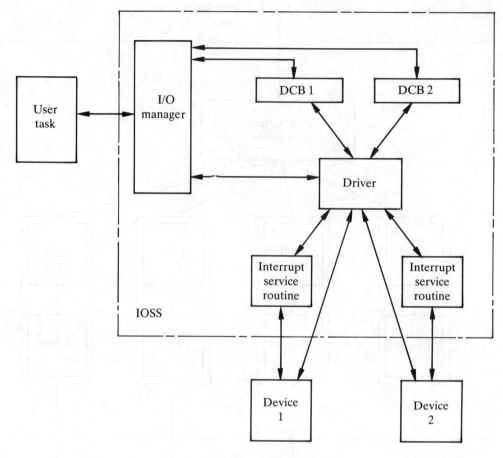

Fig. 6.30 Detailed arrangement of IOSS.

Table 6.7 Device control block

Physical device name	
Type of device	
Device address	Device-
Interrupt address	related
Interrupt service routine address	information
Device status	
Data area address	
Bytes to be transferred	Data-
Current byte count	related
Binary/ASCII	information

reasonably simple through the provision of the source code for sample DCBs which can be modified to meet particular needs.

In a multi-tasking system provision has to be made to deal with overlapping requests for a particular device, e.g., several tasks may wish to send information to the log device – typically a printer. The normal way of handling output to a printer in a single-user environment is to send a record (i.e., one line) at a time, a return being made to the user program between each line. If this is done in a multi-programming system and the printer is not allocated a specific program, there is a danger of the output from the different programs becoming intermingled. The solution usually adopted in a multi-user environment is that the output is spooled, i.e., it is intercepted by the operating system and stored in a file on the disk; when the program terminates, or the user signs off, the contents of the spool file associated with that program or that user are printed out.

A similar solution can be used in a multi-tasking environment providing that the user task can at any point force the printing-out of the spool file for that task. This addition is needed because in a real-time multi-tasking system tasks may not terminate. Although spooling provides user tasks with the ability to control interleaving of output, there is still the problem of what action to take if the device is in use when a user task makes a request to use it. There are several possible solutions:

1. Suspend the task until the device becomes available.
2. Return immediately to task with information that device is busy and leave it to the task to decide what action to take – normally to call a delay and try again later.
3. Add the request to a device request queue and return to the calling task; the calling task must check at some later time to see if the request has been completed.

There are advantages and disadvantages to each method and a good operating system will provide the programmer with a choice of actions, although not all options will be available for every device.

Option 1 is often referred to as a non-buffered request in that the user task and the device have to rendezvous; in some ways it can be thought of as the equivalent of hardware handshaking – the user task asks the device 'Are you ready?' and waits for a reply from the device before proceeding.

Option 2 is the equivalent of polling and is rarely used.

Option 3 is often referred to as a buffered request. It is a form of message passing: the user task passes to the IOSS the equivalent of a letter – this consists of both the message and instructions about the destination of the message – and the user task then continues on the assumption that eventually the message will be delivered, i.e., sent to the output device. Usually some mechanism is provided which enables the user task to check if the message has been received, i.e., a form of recorded delivery in which the IOSS records that the message has been delivered and allows the user task to check. Buffered input is slightly different in that the user task invites an external device to send it a message – this can be considered as the equivalent of providing your address to a person or to a group of people. The IOSS will collect the message and deliver it, but it is up to the user task to check its 'mail box' to see if a message has been delivered.

6.9.1 Example of an IOSS

The description which follows is of a particular IOSS of an RTOS which supports both computer peripherals – VDUs, printers, disk drives, etc. – and process-related peripherals – analog and digital input and output devices. The system commands used to access the IOSS functions are listed in Table 6.8. In addition to the commands listed in the table there are commands for analog output, for pulse output devices and for incremental output devices.

Table 6.8   IOSS system commands for RTOS

DTRC01	Disk transfer request – buffered.
DTRC02	Disk transfer request – non-buffered.
DTRC03	Call to check for completion of buffered request.
INRC01	Input request from keyboard device – buffered.
INRC02	Call to check for completion of input request.
OUCC01	Call to (a) request system data area, i.e., spool area; (b) request user data area; (c) check status of device; and (d) check if user data area is free.
OURC01	Request output of message to printer or terminal – buffered.
FMRC01	Find and reserve area of memory external to the calling task.
RMRC01	Release area of memory found using an FMRC01 call.
SCRC10	Check if the previously requested buffered scans have been completed.
SCRC11	Request non-buffered, non-priority analog scan.
SCRC12	Request non-buffered, priority scan.
SCRC13	Request buffered, non-priority scan.
SCRC14	Request buffered, priority scan.
DORC01	Request for a normal digital output, non-priority.
DORC02	Request for a normal digital output, priority.
DORC03	Request for a timed digital output, non-priority.
DORC04	Request for a timed digital output, priority. (Note: all the DORC requests can be buffered or non-buffered – the selection is made by setting a parameter for the call.)

The IOSS system manager maintains a device request queue for each device and is responsible for interpreting the user task request and placing the appropriate information in the device request queue. If the request is a buffered request, a return is made immediately to the calling task. If the request is non-buffered, then the IOSS manager changes the status of the calling task to LOCKED OUT and jumps to the dispatcher to begin the search for other work. The IOSS manager, in addition to dealing with requests, has to take action on the completion of a transfer. The driver associated with a given device signals the IOSS manager on completion of a transfer. For non-buffered requests, the IOSS manager sets the status of the user task which made the request to ON. For buffered requests there are two possible actions: if the calling task has checked to see if the action has been completed before it was completed it will have been placed in the LOCKED OUT state and hence the IOSS treats it as a non-buffered request; but if completion occurs prior to a check for completion by the user task then the IOSS records that the transfer has been completed and, when a check is made, a return to the calling task will be made with an indication that the transfer is complete. The actual detail of the actions on completion varies for the different types of device.

In addition to dealing with the above the IOSS manager, following completion of a transfer by a device, has to check if further requests are waiting in the device request queue; if they are it transfers information to the DCB and initiates the start of transfer before returning to the dispatcher.

6.9.2 Output to printing devices

The RTOS provides the programmer with a choice of spooling mechanisms:

1. System data areas   A number of fixed sized areas on the disk are provided; these are identified by a tag number.
2. User data areas   The user may define data areas of any size; these can be in memory or on the disk and are again identified by a tag number.

The system data areas are made available to any user task which requests a data area; the request can be only for a system data area, not one with a specified tag number. A user data area can be assigned to a particular user task. (Note that in this system the assignment is by implication only; all user tasks have to agree that a given tag number applies to a given data area and have to agree that use will be restricted to a given task.) The advantages of a user data area are:

1. The area can be in memory or on disk.
2. The area can be of any size.
3. If use of the area is restricted to one task then it can contain a mixture of permanent and variable data and the user task only needs to transfer the information which has been changed since the last output.

The sequence of operations to be carried out in order to output a message via a user data area is as follows:

```
; request for user data area
;
        LDA    label
        SPB    OUCC01     ;system call
                          ;return here if area in use
                          ;normal return, request accepted
;
label   DW   parameters ; type of request
                         ; tag number
                         ; device check yes/no
                         ; device number
;
;  transfer of data to data area can take place
        ........
;
;  request for output to device
;
        SPB   OURC01      ;system call
        LDK   labela
                          ;return is made to this location
;
labela DW parameters  ;device number
                      ;data area type
                      ;address of start of data area
;
;further processing can be done during data output
;
        ........
;
;check for completion of transfer
;
        LDA   label
        SPB   OUCC01
                          ;return here if not complete
                          ;return here if complete
```

6.9.3 Example of input from keyboard

For input the system expects the user task to provide a buffer area in memory to contain the input. The size is limited to a maximum of 256 words and hence an input record from the keyboard, including control characters used to edit the input line, is restricted to 256 characters. The input buffer area can either be part of the user task, or a separate area of memory found using the FMRCØ1 call. In this particular RTOS it is preferable to use a system area provided by the FMRCØ1 call, since this allows the user task to be swapped out of memory during the input.

The steps involved in the input request are shown on the flow chart in Figure 6.31 and this also shows the different layers of operation of the operating system. In outline the steps are:

1. Use FMRCØ1 call to obtain input buffer area.
2. Request input using INRCØ1.

Input/Output Sub-system (IOSS)

3. Do other processing if required.
4. Check if input completed using the INRCØ2 call – if input is not complete the task will be suspended until it is.
5. Transfer input from buffer area to program area.
6. Release buffer area using RMRCØ1 call.

It should be noted that the operating system treats the user task differently when it checks for completion of input compared with the check for completion of output. The reason for this is that it is assumed that a check for completion of input will only be made when there is no other work for the task to do; hence if a return to

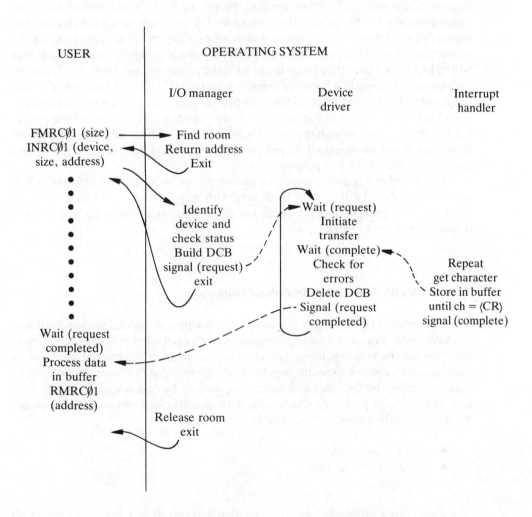

Fig. 6.31 IOSS operations for input of data.

the task on non-completion was made then all the task could do would be to delay for a short period and then check again: an inefficient procedure. On output, however, a task which finds that the previous output is not complete may be able to take some other action, e.g., set up another data area with a further request and continue.

6.9.4 Device queues and priorities

In a real-time system a simple device queue based on a first-in-first-out organization can cause problems in that the task requesting a device effectively loses its priority. Figures 6.32a and 6.32b illustrate this. In Figure 6.32a a number of tasks are queued waiting for the printer and one task (76) is already using the device. The higher priority tasks including the very high priority Task 5 will not gain access to the printer until Task 76 releases it. If the tasks have made a non-buffered request they will be locked out until they reach the head of the printer queue. However, if Task 5, for example, has made a buffered request it will be able to continue and, if it runs frequently after a short period of time, the printer queue will contain several requests from Task 5 as is shown in Figure 6.32b. If the system is not overloaded the printer will eventually catch up with the output from Task 5. The delay between the requests from Task 5 and the eventual output on the printer could be reduced if the printer queue were organized on a priority basis.

The position regarding priorities becomes even more complicated when it is remembered that the IOSS will be dealing with many devices and hence will have several device queues. Some decision has to be made as to the order in which the device queues are serviced.

6.10 TASK COOPERATION AND COMMUNICATION

The tasks which are running in the system are designed to fulfill a common purpose and to do so they must be able to communicate with each other in order to cooperate; they may also be in competition, however, for resources of the computer system – a particular data area or a specific input/output device – and this competition must be regulated either by the inter-task communication or by operating system intervention. Some of the problems which arise have already been met in considering the input/output sub-system and they involve:

- mutual exclusion;
- synchronization; and
- communication.

These three problems have been studied extensively in a branch of computer science which deals with *concurrent programming* which is the name given to the

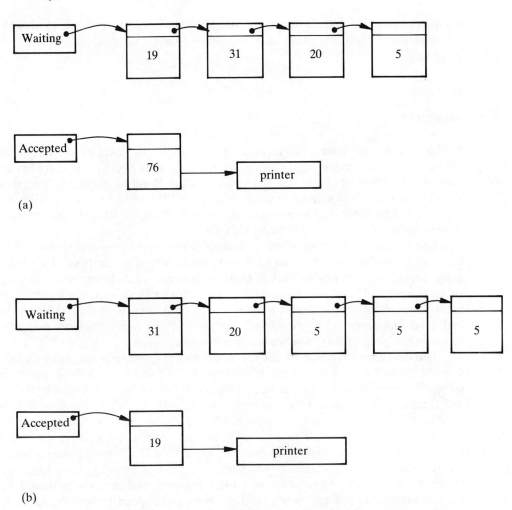

Fig. 6.32 Printer queue. (a) Buffered request; (b) non-buffered request.

programming notations and techniques for expressing potential parallelism and for solving the resulting synchronization and communication problems. Concurrent programming grew out of studies of operating systems but it has a more general applicability and is dealt with in detail in the next chapter.

The traditional operating systems have only very limited mechanisms for synchronization and communication. Synchronization mechanisms, if provided at all, are limited to signals based on the use of semaphores; such techniques are described in detail in Chapter 7. Communication mechanisms are typically the use of a SYSTEM COMMON area which is made available to all tasks and message passing via the operating system. In some operating systems it may be possible to allocate different access rights for the system common – read, write, read and write – to a given task.

6.11 SUMMARY

In this chapter we have concentrated on describing the features of traditional operating systems. Such operating systems are usually specific to a particular computer or range of computers. Examples are the Digital Equipment Company RT/11 and RSX/11 operating systems for the PDP11 series, the Data General RTOS and RDOS for the Nova range, and more recently the RMX-80 for the Intel 8080 range and OS-9 for the Motorola 6809 series.

The advantage of many of the traditional operating systems is their wide user base and the fact that they have seen extensive use in control applications. There has, however, been a tendency for the size of the operating systems to increase with each successive upgrade and it is often difficult to create small sub-sets for a particular application. Another disadvantage with many is that access to the system from high-level languages is very restricted and the addition of new devices normally requires hardware drivers to be written in assembler.

The development of the MASCOT environment represents one way in which some of the problems of lack of standardization and difficulty of accessing operating system functions have been addressed. A similar approach has been taken by Baker and Scallon [1986] with the Rex architecture. As with MASCOT the Rex system presents the user with a virtual machine which hides the details of the operating system and the hardware. The detailed procedures required for carrying out the various functions of the application program are written in a conventional language and compiled using a standard compiler. A separate language and language assembler is provided to describe how the components of the system should be connected together to form a multi-tasking system and to describe how the data sets can be shared. At this stage decisions on the number of processors to be used are made.

The system has been designed for use in the aerospace industry and the problem of overheads involved in context switching has been carefully considered. Individual processes are short procedures which once started are not interrupted: they are considered to be the equivalent of a single assembler instruction. Allocation of storage for data and code for processes is static.

An alternative to the MASCOT and Rex approach is to make available in the high-level application programming language facilities to support the creation, scheduling and synchronization of tasks. This approach is considered in the next chapter.

EXERCISES

6.1 Draw up a list of functions that you would expect to find in a real-time operating system. Identify the functions which are essential for a real-time system.

6.2 The system described in Sections 5.9 and 5.10 of Chapter 5 is to be implemented using an operating system with the task state commands listed in Table 6.6. Decide on a task priority allocation scheme and explain the reasons for your choice of priorities. Show how the timing of the various tasks could be implemented.

6.3 Discuss the advantages and disadvantages of using
  (a) fixed table; and
  (b) linked list
methods for holding task descriptors in a multi-tasking real-time operating system.

6.4 A range of real-time operating systems are available with different memory allocation strategies. The strategies range from permanently memory-resident tasks with no task swapping to fully dynamic memory allocation. Discuss the advantages and disadvantages of each type of strategy and give examples of applications for which each is most suited.

REFERENCES AND BIBLIOGRAPHY

ALLWORTH, S.T. (1981), *Introduction to Real-time Software Design*, Macmillan

BARNEY, G.C. (1985), *Intelligent Instrumentation*, Prentice Hall

BROOKES, G., MANSON, G.A. and THOMPSON, J.A. (1985), *CP/M 80 System Programming*, Blackwell

BUDGEN, D. (1985), 'Combining MASCOT with Modula-2 to aid the engineering of real-time systems', *Software: Practice and Experience*, **15(8)**, pp. 767–93

CASSELL, D.A. (1983), *Microcomputers and Modern Control Engineering*, Reston Publishing Co

COMER, D. (1985), *Operating Systems Design*, Prentice Hall

IEEE (1985), *IEEE Trial-use Standard Specifications for Microprocessor Operating Systems Interfaces*, Wiley

JOHNSON, C.D. (1984), *Microprocessor-based Process Control*, Prentice-Hall

KAISLER, S.H. (1982), *The Design of Operating Systems for Small Computer Systems*, Wiley

KOIVO, H.N. and PELTOMAA, A. (1984), 'Microcomputer real-time multi-tasking operating systems in control applications', *Computers in Industry*, **5**, pp. 31–9

LISTER, A.M. (1979), *Fundamentals of Operating Systems*, Macmillan (2nd Edition)

MELLICHAMP, D.A. (ed.) (1983), *Real-time Computing with Applications to Data Acquisition and Control*, Van Nostrand

MONTAGUE, G.A., 'Control Modules for the RML 380Z', MEng. dissertation, Control Engineering, University of Sheffield

TZAFESTAS, S.G. (ed.) (1983), *Microprocessors in Signal Processing, Measurement and Control*, Reidel, Dordrecht

7

Concurrent Programming

7.1 INTRODUCTION

In the traditional operating systems described in the previous chapter it is assumed that access by an application program to the machine hardware and the various software routines is through a limited number of operating system calls and language constructs. As a consequence many traditional operating systems have become very large and unwieldy as the designers have tried to provide facilities to meet every need. However, few applications require all the facilities provided.

With the development of microprocessors, and in particular their use as embedded controllers in which the code is held in ROM, the need has arisen for modular operating systems in which only the parts which are required for a particular application are stored in the ROM.

The use of modules, or software components, extends beyond operating system components into the idea of applications components being assembled to form the overall system. The reasons for the approach are simple – economics; a large part of the cost of any application involving a computer system is the cost of the software. If software components can be used for several applications then the unit cost for each application can be reduced. The idea of software components can be taken further with the software module being programmed into ROM and if that particular module is required then the appropriate ROM chip is added to the system in the same way in which a peripheral controller chip may be added.

The difficulty of the software component or module approach being applied to operating systems is that it is a fundamental requirement of the operating system that the various components cooperate with each other. There is a point at which the operating system cannot be subdivided further into components and still maintain the necessary degree of cooperation and coordination: there has to be a nucleus or kernel to which the components are attached. The operating system can be visualized as shown in Figure 7.1.

By analogy with hardware, the software modules can be considered as being plugged into a *software bus*. Just as the hardware bus structure has specific electrical, functional and synchronization requirements, the software bus also has specific data, functional and synchronization requirements which must be satisfied. Application tasks can be connected either directly to the software bus or through a variety of different operating system modules. That is, the system designer can choose, as with the hardware, to use a standard component, in this case a particular

Concurrent Programming

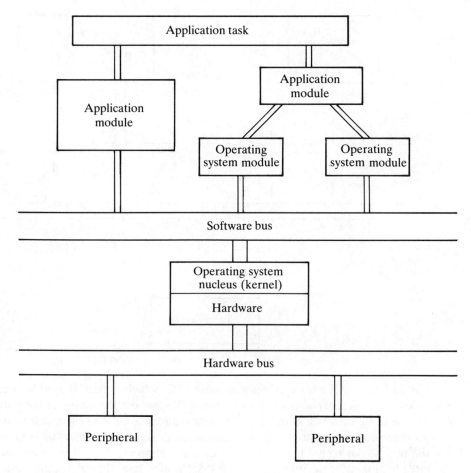

Fig. 7.1 General structure of system showing hardware and software components.

software module, and interface the application code to that module, or she/he can choose to interface the application code directly to the software bus. The nucleus or kernel can be considered to be the bus controller.

For real-time applications the important questions regarding the specification of the software bus are concerned with mutual exclusion, synchronization and communication and these operations have been extensively studied by computer scientists under the heading 'concurrent programming'.

7.2 CONCURRENT PROGRAMMING

In concurrent programming it is assumed that there is a processor available for each task: no assumption is made as to whether the processor will be an independent unit

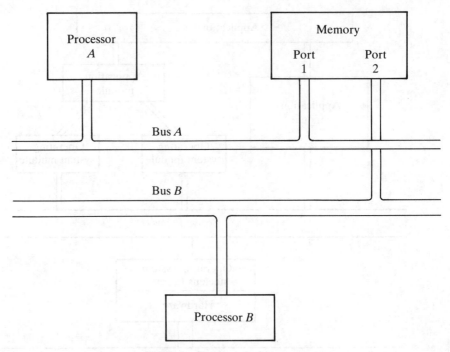

Fig. 7.2 Dual ported memory.

(i.e., a separate CPU) or simply a single CPU which is time-shared between the tasks. Furthermore, no assumption is made either about the timing or the priority of any task: tasks which require to exchange information or synchronize for other purposes do so by means of *signals*; this can be thought of as somewhat equivalent to the handshake technique used in hardware interfaces. As will be seen later, the introduction of priorities in the task scheduler can simplify some concurrent programming problems.

Another assumption which is made is that a number of so-called *primitive instructions* exist. These are instructions which are part of a programming language or the operating system and their implementation and correctness is guaranteed by the system. All that is of concern to the user is that an accurate description of the syntax and semantics is made available. In practice, with some understanding of the computer system, it should not be difficult to implement the primitive instructions.

Underlying the implementation will be an eventual reliance on the system hardware. For example, in a common memory system there will be some form of *arbiter* which will provide for mutual exclusion in accessing an individual memory location. The arbitration mechanism for a common bus structure was discussed in Chapter 3 in which the CPU controls access to the bus. If, for example, direct memory access is used then the hardware associated with the DMA unit has to inform the CPU when it wants to take control of the bus. Some systems have been designed with memory shared between processors, each of which has its own bus; in these cases dual ported memory devices have been used (see Figure 7.2) and the problem of mutual exclusion is thereby transferred to the memory itself. (Note that true dual ported memory allows concurrent access to both processors. Often the memory does not allow true concurrent access: it delays one device for a short period of time. However, the memory appears, to the processor, to permit concurrent access.)

In studying the problems of concurrency it helps if as many of the details of the actual applications as possible can be ignored. As a consequence a number of problems representing abstractions of widely occurring problems are studied in concurrent programming. In the sections which follow abstract problems will be related to some of the real problems and the various solutions to the abstract problems will be studied.

7.3 MUTUAL EXCLUSION

One of the principles of a multi-tasking system is that the resources of the system can be shared between the several tasks. But this does not imply that the resources can be used simultaneously, or that the allocation of a resource to a specific task can be made at any time. In Chapter 3, in discussing interrupts, it was noted that an interrupt can be acknowledged only at the end of an instruction cycle not in the middle of such a cycle. The instruction forms a critical section which must not be interrupted; the implication of this is that the CPU can be used only by one instruction at any one time. At the task level similar constraints apply and some resources are restricted to being used by only one task at a time; for others, such as re-entrant code segments for example, the CPU can be used by more than one task. The restriction of use to one task at a time arises for one of two reasons.

1. The nature of the resource is such that it cannot sensibly be shared by two or more tasks. This applies typically to physical devices: it does not make sense to allow a printer to be shared such that allocation to tasks can change from character to character being output.
2. The resource is such that if it is accessed by several tasks concurrently the action of any one task could interfere with the actions of another task.

* * *

Example 7.1

Consider the transfer of information from an input task to, say, a control task as is shown in Figure 7.3. The input task is assumed to get the values for three variables *KP, KI* and *KD*, representing the controller parameters and these are to be transferred to the CONTROL task. A simple method is to hold the parameter values in an area of memory which has been declared as being COMMON and hence is accessible to both tasks. Unless the input task is given exclusive rights to this COMMON data area while it writes the parameter values, there is the danger that the control task will read one new value, say, *KP* and two old values *KD* and *KI*.

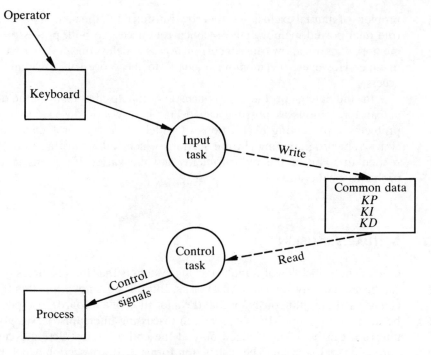

Fig. 7.3 Data sharing using common memory.

Giving exclusive rights to the input task is not a satisfactory solution in this case as will be seen later.

* * *

Example 7.2

As another example of the need for mutual exclusion let us consider the following problem. As part of the maintenance procedures a record is kept of accesses to a particular device (after a specified number of accesses some preventative maintenance has to be carried out). The system designer arranges that each task in the system which uses the device will increment a common variable, deviceUse, by using the code

    deviceUse := deviceUse + 1

The hardware will resolve the problem of simultaneous access to the memory location in which deviceUse is stored, but this is not sufficient to guarantee the correct functioning of the counter.

Consider the following scenario. The compiler will generate machine code in the following form:

    load deviceUse
    add 1
    store deviceUse

Suppose the value of deviceUse is 38 and Task A executes the load deviceUse, then the tasks reschedule and Task B executes load deviceUse: both tasks now have in their own environments a register containing the current value of deviceUse, i.e., 38. Task B now executes the 'add 1' instruction and the 'store deviceUse' instruction giving a value of 39 in deviceUse. Control is now returned to Task A which executes the 'add 1' and the 'store deviceUse' instructions which again give a value of 39. The final value of deviceUse is thus 39 even though it started as 38 and has been incremented twice.

* * *

The abstraction of the mutual exclusion problems given above can be expressed in the form

remainder 1
pre-protocol
critical section
post-protocol
remainder 2

where remainder 1 and remainder 2 represent sequential code which does not require access to a particular resource or to a common area of memory. The critical section is part of the code which must be protected from interference from another task. The 'protocols' called before and after the critical sections are code which will ensure that the critical section is executed so as to exclude all other tasks. In order to benefit from concurrency both the critical section and the protocols must be short such that the remainders represent a significant body of code which can be overlapped with other tasks. The protocols represent an overhead which has to be paid in order to obtain concurrency.

It is implicit in concurrent programming that there is 'loose connection' or 'low coupling' between tasks (see, e.g., Pressman for definition of 'loose connection'). The low coupling can increase reliability in that a bug in one task which causes an abnormal termination of that task will not, if there is low coupling, cause other tasks in the system to fail. In abstract terms this can be expressed as the requirement that an abnormal termination in the code forming the remainder should not affect any other task. It would be unreasonable to demand that a failure of the protocol or the critical sections did not affect another task, since the critical section represents the code by which communication or sharing of a resource with another task is taking place.

7.3.1 Primitives

In considering solutions to the abstract problems involving mutual exclusion, etc., a number of so-called primitive instructions will be introduced: these typically form the protocols. It is assumed that any implementation of these primitives as part of a language or as part of an operating system will guarantee their correctness and that all which need concern us is the provision of an accurate description of their semantics and syntax. It is a basic assumption that a given primitive forms an indivisible instruction and hence the task which invokes a primitive is guaranteed that it will not be swapped during the execution of the primitive.

In practice, it should not be difficult, with some understanding of the hardware of the computer system, to implement the primitives. It should be noted that underlying the implementation of the primitives and hence mutual exclusion is an eventual reliance on the system hardware and the guarantee of exclusive access at the hardware level to a specific memory location.

7.3.2 Condition flags

A simple method of indicating if a resource is being used or not is to have associated with that resource a flag variable which can be set to TRUE or FALSE (or to 0 or 1, or SET or RESET). A task wishing to access the resource has to test the flag before using the resource. If the flag is FALSE (0 or RESET) then the resource is available and the task sets the flag TRUE (1 or SET) and uses the resource. The procedure is illustrated below.

* * *

Example 7.3

```
MODULE MutualExclusion1;
(* Mutual exclusion problem Condition Flag solution 1*)
VAR
     deviceInUse: BOOLEAN;
  PROCEDURE Task; (* task assumed to be running in parallel with
                     other tasks *)
  BEGIN
  (*     remainder1   *)
     WHILE deviceInUse DO
        (*test and wait until available *)
     END (*while*)
     deviceInUse:= TRUE;   (*claim resource*)
     (*......
     use the resource - critical section
     ...... *)
     deviceInUse := FALSE;
  (*     remainder2 *)
  END Task;
(* main program *)
END MutualExclusion1.
```

* * *

In the above solution there are two problems.

1. The WHILE statement forms a 'busy wait' operation which relies on a preemptive interrupt to escape from the loop. If the task which has already

claimed the resource cannot interrupt the busy wait, the system is deadlocked as the resource will never become available. This is similar to the polling of devices described in Chapter 3.
2. The testing and setting of the flag are separate operations, hence the task could be preempted and lose its running status between checking the flag and setting the resource unavailable. A consequence could be that, as is shown in Figure 7.4, two tasks could both claim the same resource.

The two Tasks A and B shown in Figure 7.4 both share a printer and it is assumed that access to the printer is controlled by the FLAG variable printerInUse, which is set to 1 when the printer is in use and to 0 when it is available. Task A checks the printerInUse flag and finds that the printer is available, but before it can execute the next instruction which would be to set the printerInUse flag to 1 and hence claim the printer, the dispatcher forces a task status change and Task B runs. Task B also wishes to use the printer and checks the flag, it finds that the printer is available, sets the printerInUse flag to 1 and begins to use the printer. At some time later it requires some other resource and the dispatcher suspends it and makes Task A the active task. Task A now claims the printer and begins to use it. Thus both tasks think that they have the exclusive use of the printer, whereas they are both using it and the output from the two tasks will be mixed up. After some time Task A is again suspended and Task B continues. It now finishes with the printer and releases it by setting printerInUse to 0; this makes the printer available to any other task even though Task A still thinks that it has exclusive use of the printer. At task change 4 Task A again uses the printer and eventually releases it although it has in fact already been released by Task B.

For a condition flag to work securely it is therefore vital that the operations of test condition/set condition are indivisible. The importance of this has been recognized by some CPU designers by the provision of single instructions which test and set or clear a flag. In the absence of such instructions the test/set operation can be made indivisible by the use of the enable/disable interrupt instructions.

* * *

Example 7.4

```
(* Mutual exclusion problem Condition flag solution 2*)
VAR
     deviceInUse:BOOLEAN;
     deviceClaimed: BOOLEAN;
PROCEDURE Task;
BEGIN
     (* remainder1 *)
     REPEAT
         DisableInterrupts; (* procedure *)
         IF deviceInUse THEN
                 deviceClaimed:= FALSE
         ELSE
                 deviceInUse:= TRUE;
                 deviceClaimed:= TRUE
         END (* IF *)
```

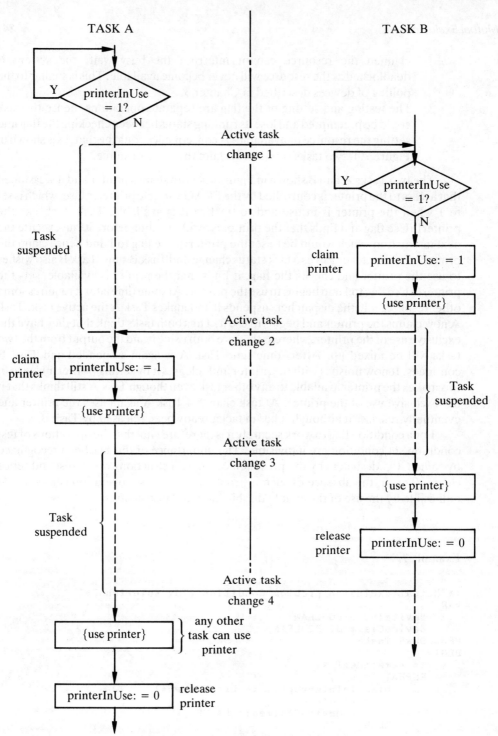

Fig. 7.4 Attempt at mutual exclusion using condition flags.

Mutual Exclusion

```
            EnableInterrupts;
    UNTIL deviceClaimed;
    (*......
    use resource (*critical section*)
    ..... *)
    DisableInterrupts;
    deviceInUse:=FALSE;
    deviceClaimed:= FALSE;
    enable interrupts;
    (* remainder2 *)
END Task;
```

*   *   *

Solution 2 is an improvement in that it will prevent two tasks gaining access to the same resource. It still has the problem of being in an endless loop waiting for the resource to become available, and it relies on some form of preemption to allow other tasks to run. If this approach is used it would be sensible in practice to incorporate a request for a short delay between each testing of the condition flag. The insertion of a call delay statement between lines 13 and 14, i.e.

```
    enable interrupts;
    delay(delayTime);
    UNTIL deviceClaimed;
```

would be appropriate.

Because errors in the interrupt enable/disable status are potentially dangerous (a failure to enable interrupts at the end of a critical section will cause the whole system to fail), manipulation of the interrupt status flag should be restricted as much as possible and preferably should not occur in application level programs; the above solution is therefore not recommended.

7.3.3 Semaphores

A solution to this problem was provided by Dijkstra in 1968 with the introduction of the concept of the binary semaphore. A binary semaphore is a condition flag which records whether or not a resource is available. If, for a binary semaphore s, $s = 1$ then the resource is available and the task may proceed, if $s = 0$ then the resource is unavailable and the task must wait. In order to avoid the processor wasting time semaphores are usually implemented with a task or condition queue to which the waiting process is appended. There are only three permissible operations on a semaphore: initialize, secure, and release. A binary semaphore can be defined as a condition flag variable with an associated task queue and the following operations:

Initialize (s:ABinarySemaphore, v:INTEGER): set semaphore s to value of v ($v = 0$ or 1)

Secure(s): if $s = 1$ then $s := 0$ else suspend the calling task and place it in condition queue(s)

Release(s): if condition queue(s) is empty then $s := 1$ else resume first task in condition queue(s).

The operations secure(s) and release(s) are system primitives which are carried out as indivisible operations and hence the testing and setting of the condition flag are performed effectively as one operation.

As an example consider a task which wishes to access a printer.

* * *

Example 7.5
```
(* Mutual exclusion problem solution 3 - use of binary semaphore*)
VAR
      printerAccess: SEMPAPHORE;
PROCEDURE Task;
BEGIN
      (* remainder1 *)
      Secure(printerAccess)   (*if printer is not available task
                                 will be suspended at this point*)
      (*printer available - critical section*)
      (* .......
      do output
      ...... *)
      Release(printerAccess)
      (* remainder2 *)
END Task ;
```

In Figure 7.5 the underlying operations which take place as several tasks attempt to access the same resource (assumed to be a printer) are shown. The binary semaphore printerAccess is initialized to the value 1 in Step 1; as part of this process a three-item record with the semaphore value set to 1 and the pointers to the head and tail of the semaphore queue set to null is created. In Step 2, Task A performs a Secure(printerAccess) operation and the semaphore value is set to 0. Since there was no other task waiting, Task A is allowed to use the resource. Sometime later Task A suspends and Task B performs a Secure(printerAccess), Step 3; since printerAccess = 0 it cannot continue and is added to the semaphore queue by inserting, in the semaphore control block, head and tail pointers to the task descriptor for Task B. If Task C now performs a Secure(printerAccess), Step 4, then the pointer in the task descriptor for Task B is filled in with the address of the TD for Task C and the tail pointer in the semaphore control block is filled in to point to Task C. When Task A performs the Release(printerAccess) operation, Step 5, Task B is removed from the semaphore queue and then obtains access to the resource. At Step 6, Task B performs the Release(printerAccess) operation and hence Task C is allowed to run and at Step 7 when Task C performs Release(printerAccess) the value of the semaphore is set to 1.

In the above arrangement it is assumed that the tasks gain access to the resource in the order in which they performed the secure(s) operation, i.e., the task which has been waiting longest is served first. It is possible to organize the semaphore queue so that the tasks are queued according to their priority.

* * *

1. Initialize (printerAccess, 1)

printerAccess

Value = 1
Head
Tail

2. Task A active secure (printerAccess)

printerAccess

Value = 0
Head
Tail

3. Task A suspends, Task B active secure (printerAccess)

printerAccess

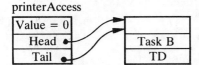

Task B is suspended and placed in printerAccess queue

4. Task C runs attempts secure (printerAccess)

printerAccess

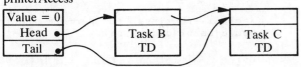

Task C is suspended and added to printerAccess queue

5. Task A runs, Release (printerAccess)

printerAccess

Task B is removed from printerAccess queue and placed in Ready queue

6. Task B runs, Release (printerAccess)

printerAccess

Value = 0
Head
Tail

Task C is transferred from printerAccess queue to Ready queue

7. Task C runs, Release (printerAccess)

printerAccess

Value = 1
Head
Tail

printer is available to any task

Fig. 7.5 Mutual exclusion using binary semaphore.

* * *

Example 7.6   Use of semaphores to solve transfer of controller parameters problem

```
(*Mutual exclusion - transfer of controller parameters solution 1*)
TYPE
    AParameterRecord = RECORD
                           kp : REAL;
                           kd : REAL;
                           ki : REAL
                       END;
VAR
    mutex : SEMAPHORE;
    controlParameters : AParameterRecord;
    inputBlock : AParameterRecord;

TASK DataTransfer;
(* transfers input data to the controller *)

BEGIN
    Secure (mutex);
    controlParameters := inputBlock;
    Release (mutex);
END DataTransfer;

TASK Control;

BEGIN
    Secure (mutex);
    DoControl              (*actual routines to perform control
                            would be placed here*)
    Release (mutex);
END Control;

BEGIN (*main body of program *)
    Initialize (mutex, 1);
    StartTask (DataTransfer);
    StartTask (Control);
END Main.
```

* * *

The above is an acceptable solution for cooperating tasks in a non-real-time environment, but for real-time work there are several problems. The first is in TASK dataTransfer, the use of a semaphore does not prevent the task being suspended and another task being run during the critical section: it prevents any other task accessing the controlParameters record (provided that the task checks the semaphore mutex before processing). If, for example, it is time to run the control task, which will have a higher priority than the dataTransfer, the control task will check the mutex semaphore and will then be suspended waiting the completion of the data transfer by the dataTransfer task. The consequences of the delay could be unpredictable: if the dataTransfer task is the only other task waiting to run, or has the highest priority of any of the waiting tasks, then the solution could be acceptable in that the transfer will be completed and the control task will run immediately the operation release (mutex) is performed. However, in the case in which, e.g., a displayUpdate task has a higher priority than the dataTransfer task, the sequence of events could be as illustrated in Figure 7.6.

Mutual Exclusion

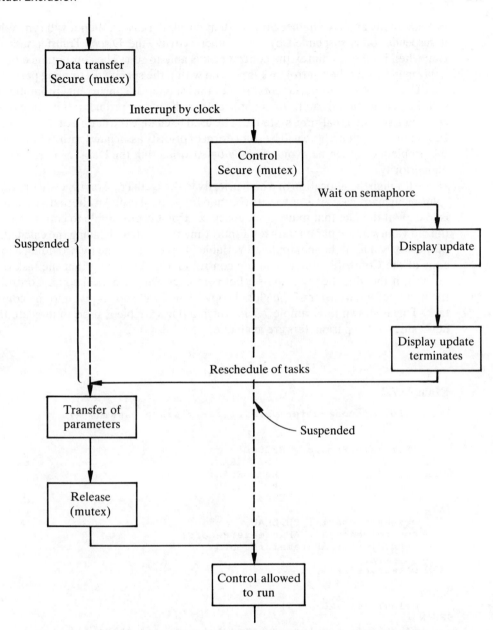

Fig. 7.6 Transfer of controller parameters – use of semaphore.

The dataTransfer task is assumed to have just secured mutex, at which time the clock interrupt forces a rescheduling of the tasks and the control task runs. The control task will be suspended when it attempts to secure mutex and there will again be a rescheduling of the tasks; if it is assumed that the displayUpdate task has now become ready and is of higher priority than DisplayTransfer, then it will run. When it suspends, or is suspended by some other activity, the DisplayTransfer task can run; when it releases mutex the control task is able to run. The consequence of this delay may be that the control task is not run within the specified sampling period.

The real-time requirements make the standard concurrent programming solution using mutual exclusion inapplicable in the above problem, but some of the characteristics of a real-time system can be used to change the nature of the problem. In a real-time system there will be some form of priority associated with certain tasks; the problem can then be avoided easily by ensuring that the DataTransfer task has high priority.

Although the allocation of a high priority to the DataTransfer task will provide a solution in many cases it is not a safe solution for general use. An alternative solution is to exploit the fact that in using feedback control it is preferable to continue using the old values of the parameters for a short time rather than delaying the calculation of the next value of the manipulated variable. This can be done by providing a local copy of the ControlParameters in the control task and ensuring that the task does not wait if the global copy is in use, but continues using the local copy. Control of the timing of the transfer of the global copy to the local copy is given to the control task. This is shown in Example 7.7 in which a flag variable is used to indicate that new values of the parameters are available.

* * *

Example 7.7

```
(*Transfer of controller parameter - message passing *)

TYPE
    AParameterRecord = RECORD
                           kp : REAL;
                           kd : REAL;
                           ki : REAL
                       END;
VAR
    messagePresent : BOOLEAN;
    controlMessage : AParameterRecord;
    inputBlock : AParameterRecord;

TASK DataTransfer;

CONST
    delayTime = 20;
BEGIN
    (* wait if previous message has not been taken *)
    WHILE messagePresent DO
        Delay(delayTime)
    END (* while *);
    controlMessage := inputBlock;
    messagePresent := TRUE;
END DataTransfer;

PROCDEURE Control;
```

```
    VAR
        controlParameter : AParameterRecord;
    BEGIN
        DoControl; (*actual control statements would go here *)
        IF messagePresent THEN
                controlParameters := controlMessage;
                messagePresent := FALSE
        END (* if *);
    END Control;
    BEGIN (*main body*)
        messagePresent := FALSE;
        StartTask(DataTransfer);
        StartTask(Control);
    END Main.
```

* * *

The above provides a satisfactory solution in that there is no danger of the control task being interrupted while it is transferring data since it is run at a higher priority than the dataTransfer task.

7.4 PRODUCER-CONSUMER PROBLEM

The producer-consumer problem is the abstraction of a widely occurring problem which arises when one task is producing data which it has to store somewhere until a second task is ready to use the data. An example of this would be an alarm logging task which was producing data for a period at a rate much greater than that at which the logging task could print it out. As is shown in Figure 7.7 input and output devices can be seen as producer and consumer tasks respectively. A buffer is used to store the data until the consuming task is ready to take it.

A buffer is an area of memory which is common to both the producing and consuming task and it should be large enough to smooth out the peaks in the rates of data transmission between the tasks. In the examples which follow it is assumed that the buffer is bounded, i.e., of finite size, and that the operation of storing an item of data in it is performed by the call

   Put(x)

and an item of data is removed by the call

   Get(x)

Since the buffer is of finite size it is necessary to know when it is full and when it is empty, the following function calls apply: full – which returns the value TRUE if the buffer is full; and empty – which returns the value FALSE if the buffer is empty.

In the first solution it is assumed that the producer and consumer are formed by separate tasks which share a common buffer area.

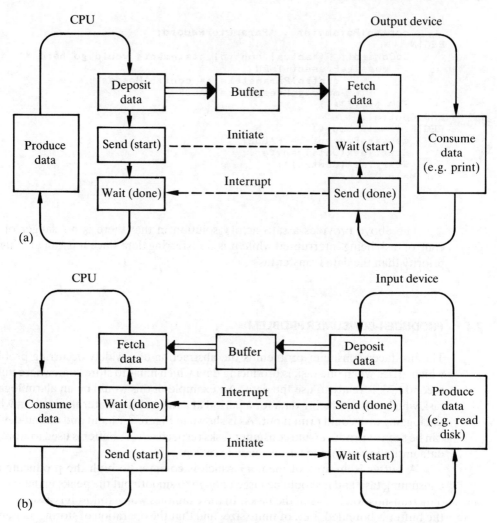

Fig. 7.7 Input (a) and output (b) device model (from Young, *Real-time Languages*, 1982).

*   *   *

Example 7.8

```
(* Producer-consumer problem - solution 1*)
VAR   commonBuffer : buffer;

TASK Producer;

VAR   x:data;
BEGIN
    LOOP
```

```
            Produce(x);
            WHILE Full DO
                Wait
            END (* while *);
            Put(x);
        END (* loop *);
END Producer;

TASK Consumer;

VAR  x:data;
BEGIN
    LOOP
            WHILE Empty DO
                Wait
            END (* while *);
            Get(x);
            Consume(x);
        END (* loop *);
END Consumer;
```

* * *

The producer operates in an endless cycle producing some item x and waiting until the buffer is not full to place x in the buffer; the consumer also operates in an endless cycle waiting until the buffer is not empty and removing item x from the buffer. The above solution is not satisfactory for two reasons:

1. The Put(x) and Get(x) are both operating on the same buffer and for security of the data simultaneous access to the buffer cannot be allowed – the mutual exclusion problem.
2. Both the producer and the consumer use a 'busy wait' in order to deal with the 'buffer full' and 'buffer empty' problem.

The first problem can be solved using the semaphore, with the operations secure and release; the second problem can be solved by using a primitive called a 'signal'.

A signal s is defined as a binary variable such that if $s = 1$ then a signal has been sent but has not yet been received. Associated with a signal is a condition queue and the permissible operations on a signal are:

Initialize (s:Signal: v:INTEGER)
    set s to the value of v (0 or 1).

Wait(s)
    if $s = 1$ then $s := 0$ else suspend the calling task and
    place it in the condition queue s.

Send(s)
    if the condition queue s is empty then $s := 1$ else
    transfer the first task in the condition queue to the
    ready queue.

It should be readily apparent that a signal is similar to a semaphore: in fact the difference between the two is not in the way in which they are implemented but in the way in which they are used. A semaphore is used to secure and release a resource and as such the calls will both be made by one task: a signal is used to synchronize the activities of two tasks and one task will issue the send and the other task the wait. Using semaphores and signals the producer-consumer problem can be solved as follows.

* * *

Example 7.9

```
(*Producer-consumer problem - solution 2*)
VAR   commonBuffer : Abuffer;
      bufferAccess : ABinarySemaphore;
      nonFull, nonEmpty : Signal;

TASK Producer;

VAR   x:data;
BEGIN
      LOOP
            Produce(x);
            Secure(bufferAccess);
            IF Full THEN
                  Release(bufferAccess);
                  Wait(nonFull);
                  Secure(bufferAccess);
            END (*if*);
            Put(x);
            Release(bufferAccess);
            Send(nonEmpty);
      END (*loop*);
END Producer;

TASK Consumer;

VAR   x:data;
BEGIN
      LOOP
            Secure(bufferAccess);
            IF Empty THEN
                  Release(bufferAccess);
                  Wait(nonEmpty);
                  Secure(bufferAccess);
            END (* if *);
            Get(x);
            Release(bufferAccess);
            Send(nonFull);
            Consume(x);
      END (*loop*);
END Consumer;
```

* * *

In the above solution it should be noted that the critical code is enclosed between secure and release operations, but it is vital that the bufferAccess semaphore is released before executing the wait(nonFull) or wait(nonEmpty) primitives. If this is not done the system will deadlock, e.g., if the producer executes wait(nonFull) while holding the access rights to the buffer then the buffer can never become non-full since the only way it can is for the consumer to remove an item of data, but the consumer cannot gain access to it until it is released by the producer.

Both semaphores and signals can be generalized to allow a semaphore or a signal variable to have any non-negative integer value – in this form they are sometimes referred to as 'counting semaphores'.

In MASCOT a slightly different approach is taken in that the functions of a semaphore and a signal are combined into one primitive – a control queue. The queue can be either free or busy and, as with a semaphore and a signal, tasks can be appended and removed from the queue. The permissible operations are defined as follows:

JOIN(controlQueue)
IF the controlQueue is free THEN it is set to busy and the priority of the calling task is set equal to the priority of the queue ELSE the calling task is appended to a list of tasks waiting to join the controlQueue.

LEAVE(controlQueue)
The priority of the calling task is set to the priority it had before executing the Join primitives; IF there are no tasks waiting to join the controlQueue THEN it is set to free ELSE a task is removed from the pending queue and added to the ready queue at the priority of the controlQueue.

STIM(controlQueue)
acts as a 'signal' on the control queue; hence IF a task is waiting THEN it is transferred to the ready queue at the priority of the controlQueue; IF there is no waiting task THEN the controlQueue is 'primed'. If multiple STIMs are received in succession they have no effect beyond the effect of the first one to be received. Any task may STIM the controlQueue.

WAIT(controlQueue)
may only be used by a task which has already joined the controlQueue. IF the controlQueue is already primed, i.e., a STIM(controlQueue) has been executed, THEN the task will continue executing ELSE it will suspend until a STIM is received.

* * *

Example 7.10

Communication between tasks in MASCOT is normally by means of *channels*. The above primitives can be used to implement a channel. It is assumed that the data being produced is a string of characters up to 80 characters long and that a single buffer of 80 characters is used in the channel. The channel can provide an interface between producer and consumer tasks.

```
(*Simple implementation of a CHANNEL*)

VAR   inQ, outQ        : AContorolQueue;
      buffer           : ARRAY [0..79] OF CHAR;
      count            : INTEGER;

TASK Producer;
VAR   data             : ARRAY [0..79] OF CHAR;
      i, noChar        : INTEGER;
BEGIN
    ...
    produce characters (noChar) and place in data
    ...
(*transfer of characters to buffer - input to CHANNEL *)
    JOIN(inQ);
    WHILE count <> 0 DO
        WAIT(inQ)  END (* while *) ;      (*wait for buffer to be emptied*)
    FOR i:= 0 TO noChar-1 DO
        buffer[i] := data[i]    (*fill buffer*)
    END (* for *);
    count := noChar-1;
    STIM(outQ);
 (*signal that information is in buffer if consumer is waiting*)
    LEAVE(inQ);
(*end of CHANNEL*)
    ...
    remainder
    ...
END Producer;

TASK Consumer;
VAR   data             : ARRAY [0..79] OF CHAR;
      i, noChar        : INTEGER;
BEGIN
(*transfer characters from CHANNEL*)
    JOIN(outQ);
    WHILE count=0 DO
       WAIT(outQ) END (* while *);     (*wait if no data in buffer*)
    FOR i := 0 TO count DO
        data[i] := buffer[i]
    END (* for *);
    noChar := count + 1;
    count := 0;
    STIM(inQ);
    LEAVE(outQ);
(*end of CHANNEL*)
    ...
    use the data
    ...
END Consumer.
```

* * *

In the above example the use of the JOIN and LEAVE operations prevents more than one task at a time attempting to take or put information into the buffer, i.e., they enforce mutual exclusion. Hence if several tasks are attempting to pass messages to the consumer task the messages will not be corrupted. It should be noted that in the above implementation the buffer can only hold one message at a time even if the message does not fill the whole buffer; a different arrangement for keeping track of the buffer pointers and the buffer empty and buffer full conditions would be required to allow for several data items to be in the buffer.

* * *

Example 7.11

Consider a system in which it is required to have a display with three windows as shown in Figure 7.8. Window 1 is used to display the normal control information about the plant; the information is supplied from a task ControlDisplay, Window 2 is used to display alarm information provided from a task AlarmDisplay and Window three is used by the OperatorInput task to display the input prompts.

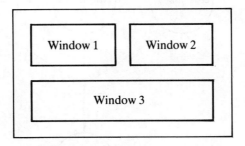

Fig. 7.8 Display requirements for Example 7.11.

One possible design solution is to use a DisplayDriver task to collect the information from the three tasks and output it to the correct window of the actual display. The tasks could be connected by a simple channel as described in Example 7.10; this arrangement is shown in Figure 7.9. It is assumed that each message will be prefixed by a code indicating its source, e.g., by letting the first two characters be 'W1', 'W2' and 'W3' to indicate Windows one, two and three respectively. The tasks ControlDisplay, AlarmDisplay and OperatorInput will all be producer tasks and will incorporate the following section of code:

```
     JOIN(inQ);
     WHILE count <> 0 DO
        WAIT(inQ)   END (* while *);   (*wait for buffer to be emptied*)
     FOR i:= 0 TO noChar-1 DO
           buffer[i] := data[i]    (*fill buffer*)
     END (* for *);
     count := noChar-1;
     STIM(outQ);
(*signal that information is in buffer if consumer is waiting*)
     LEAVE(inQ);
```

The DisplayDriver is the consumer task and will incorporate the code:
```
JOIN(outQ);
WHILE count=0 DO
 WAIT(outQ) END (* while *);      (*wait if no data in buffer*)
FOR i := 0 TO count DO
     data[i] := buffer[i]
END (* for *);
noChar := count + 1;
count := 0;
STIM(inQ);
LEAVE(outQ);
```

* * *

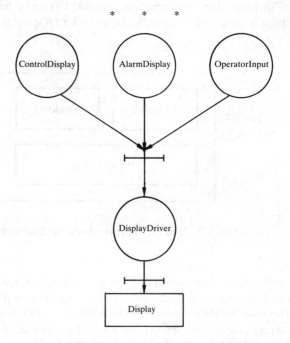

Fig. 7.9 Outline design for Example 7.11.

An important feature to note about the MASCOT primitives is that the task takes on the priority of the queue which it is accessing, hence as well as making decisions about task priorities the designer must also make decisions about the priority allocated to, for example, a channel. If, for example, the channel connecting the DisplayDriver to the other tasks in Example 7.11 is given a low priority then the AlarmDisplay task may be held up; however, if it is given a high priority, some control tasks may be delayed while the channel transmits operator input information.

Although semaphores, signals and the MASCOT primitives can be used to construct satisfactory solutions to the typical inter-task communication problems which arise, they are not easy to use and their operation is obscure, hence it is easy to make design mistakes. As an example of the sort of problems which can arise let us consider another formulation of the simple CHANNEL given in Example 7.10. At first sight it would seem that the WHILE expression DO WAIT(inQ) END; statement could be replaced with an IF expression THEN WAIT(inQ) statement, and the solution would then read:

* * *

Example 7.12

```
(*Simple implementation of a CHANNEL - incorrect solution*)

VAR   inQ, outQ         : AContorolQueue;
      buffer            : ARRAY [0..79] OF CHAR;
      count             : INTEGER;

TASK Producer;
VAR   data              : ARRAY [0..79] OF CHAR;
      i, noChar         : INTEGER;
BEGIN
    ...
    produce characters (noChar) and place in data
    ...
(*transfer of characters to buffer - input to CHANNEL *)
    JOIN(inQ);
    IF count <> 0 THEN
            WAIT(inQ)   END;       (*wait for buffer to be emptied*)
    FOR i:= 0 TO noChar-1 DO
            buffer[i] := data[i]   (*fill buffer*)
    END;
    count := noChar-1;
    STIM(outQ);
 (*signal that information is in buffer if consumer is waiting*)
    LEAVE(inQ);
(*end of CHANNEL*)
    ...
    remainder
    ...
END Producer;

TASK Consumer;
VAR   data              : ARRAY [0..79] OF CHAR;
      i, noChar         : INTEGER;
BEGIN
(*transfer characters from CHANNEL*)
    JOIN(outQ);
    IF count=0 THEN
              WAIT(outQ) END;      (*wait if no data in buffer*)
    FOR i := 0 TO count DO
            data[i] := buffer[i]
    END;
    noChar := count + 1;
    count := 0;
    STIM(inQ);
    LEAVE(outQ);
(*end of CHANNEL*)
    ...
    use the data
    ...
END Consumer;
```

* * *

If we now consider the scenario in which the producer finds the count = 0 and hence does not execute the WAIT(inQ) but proceeds to place the data in the buffer and then for some other reason suspends. The consumer task now takes the data and sets count := 0 and STIMs the inQ. The producer again finds the value of count = 0 and does not execute WAIT(inQ) but puts the data in the buffer; if the producer now instead of suspending attempts to produce the next set of data it will again test count. It now finds that count is not 0 and hence it executes WAIT(inQ), but because inQ is already primed it is allowed to continue and place data in the buffer thereby overwriting the existing data. This problem does not arise with the WHILE .. DO construct. (Explain why it does not occur.)

An alternative solution to using the WHILE .. DO construct would be to test if a task is waiting and if there are no tasks waiting then do not execute the STIM statement. However, it is not possible to test the semaphore-type primitives and hence the programming of alternative actions is not possible. It is frequently the case in real-time systems that a task would wish to continue without using a resource if a particular resource was not available (see section 7.3.2 above).

In addition to the above problems the use of these primitives involves each task in being given knowledge of the various buffers, signals, semaphores and condition variables which are used and places the responsibility of using them correctly on the application programmer.

7.5 MONITORS

An alternative solution to the use of semaphores to provide mutually exclusive access to a shared resource is the idea of a monitor which was introduced by Brinch Hansen [1973, 1975a] and by Hoare [1974]. A monitor is a set of procedures which provide access to data or to a device: the procedures are encapsulated inside a module which has the special property that only one task at a time can be actively executing a monitor procedure. It can be thought of as providing a fence around critical data, but unlike the use of condition flags or semaphores, the operations which can be performed on the data are moved inside the fence as well as the data itself.

An example of a simple monitor is given in Figure 7.10. Two procedures are provided to access the data: WriteData and ReadData; these procedures represent 'gates' through which access to the monitor is obtained. The monitor prevents any other form of access to the critical data. If a task wishes to write data it calls the procedure WriteData and as long as no other task is already accessing the monitor it will be allowed to enter and write new data. If any other task was already using either the WriteData or ReadDate operations then the task would be halted at the gate and suspended, since only one task at a time is allowed to be within the monitor fence.

A more complicated monitor is shown in Figure 7.11 in which there are three entry points: Entry 1, Entry 2 and Entry 3 and two conditions on which tasks which have gained entry may have to wait. In the figure, one task, T15, is in the monitor and

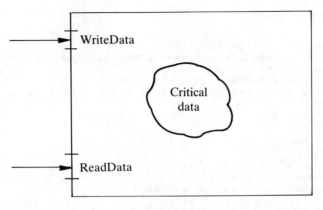

Fig. 7.10 Simple monitor.

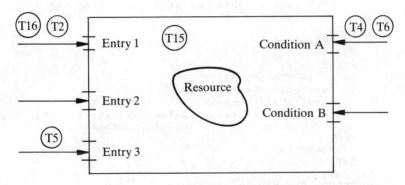

Fig. 7.11 General monitor.

three tasks are waiting to enter, two at Entry Point 1–T16 and T2 – and one at entry point three – T5. Two tasks have previously entered and have been suspended waiting for Condition A. There are no tasks waiting at Entry Point 2 or for Condition B.

The advantage of a monitor over the use of semaphores or other mechanisms to enforce mutual exclusion is that the exclusion is implicit: the only action required by the programmer of the task requiring to use the resource is to invoke the entry to the monitor. If the monitor is correctly coded then an applications program cannot use a resource protected by a monitor incorrectly. Monitors themselves do not provide a mechanism for synchronizing tasks and hence for this purpose the monitor construct has to be supplemented by allowing, e.g., signals to be used within it. An example of a monitor which uses signals is shown below; this is based on using Modula 2.

* * *

Example 7.13

```
MODULE Buffer[monitorPriority];
(*solution to producer-consumer problem, also implementation of a
simple CHANNEL *)
   FROM Signals IMPORT
           Signal, InitSignal, AwaitSignal, SendSignal;
   EXPORT
           Put, Get;
   CONST
           nMax = 32;
   VAR
     nFree, nTaken: [0..nMax];
     in, out: [1..nMax];
     b: ARRAY [1..nMax] OF INTEGER;
     notFull, notEmpty: Signal;
   PROCEDURE Put(i: INTEGER);
   BEGIN
      IF nFree = 0 THEN
         AwaitSignal(nonFull)
            (* another task can call Put during the wait - it will
               also find nFree=0 and will wait*)
      ELSE
         DEC(nFree)
      END;
      b[in] := i;
      in := in MOD nMax + 1;
      IF Awaited(nonEmpty) THEN
         SendSignal(nonEmpty)
            (* a higher priority task waiting for this signal will
               run now and may lead to another call of Put*)
      ELSE
         INC(nTaken)
      END
   END Put;
   PROCEDURE Get(VAR i: INTEGER);
   BEGIN
      IF nTaken = 0 THEN
         AwaitSignal(nonEmpty)
            (* another task can call Get during the wait - it will
               also find nTaken = 0 *)
      ELSE
         DEC(nTaken)
      END;
      i := b[out];
      out := out MOD nMax + 1;
      IF Awaited(nonFull) THEN
         SendSignal(nonFull);
            (* a higher priority task waiting for this signal will run
               now and may lead to another call of Get*)
      ELSE
         INC(nFree)
      END
   END Get;
BEGIN
   nFree := nMax; nTaken := 0;
   in := 1; out := 1;
   InitSignal(nonFull); InitSignal(nonEmpty);
END Buffer;
```

If the use of signals to allow for task synchronization is allowed within a monitor then, in order to prevent deadlock, a task which gains access to a monitor procedure but then executes a WaitSignal operation must be suspended and placed outside the monitor in order to allow another task to enter – hence in Figure 7.11 the tasks waiting on conditions are shown outside the monitor. In the producer-consumer problem shown above, suppose the producer task enters the monitor with a call to Put, but is forced to wait because the buffer is full, then the only way in which the buffer can become non-full is if another task, the consumer, is able to enter the monitor and remove an item from the buffer using the Get procedure. The consumer will then issue a SendSignal to awaken the producer task, unless the SendSignal operation is the last executable statement in the Get procedure then two tasks could be active within the monitor, thus breaching the mutual exclusivity rule. Hence the use of signals within a monitor requires the rule that the Send operation must be the last executable statement of a monitor procedure.

The RTOS operating described earlier as an example of a traditional operating system is constructed as a monolithic monitor in that all the functions, or all the resources, are contained within the one monitor; the modern practice is to provide an ability to construct monitors for each resource and to provide a means by which the final system includes only those monitors for which there is a need.

For real-time control the standard monitor construction as outlined above has a similar limitation to the semaphore in that it does not reflect the priority of the task trying to use a resource: the first task to gain entry can lock out other tasks until it completes and hence a lower priority task could hold up a higher priority task in the manner described in Example 7.6. In the traditional operating systems this was avoided by ensuring that once an operating system call was made (in other words when a monitor function was invoked), the call would be completed without interruption from other tasks, i.e., the monitor function is treated as a critical section. It should be noted that this does not mean that the whole operation requested was necessarily completed without interruption; thus, for example, a request for access to a printer for output would be accepted and the request queued. Once this had been done another task could enter the monitor to request output and either be queued, or receive information from the monitor as to the status of the resource. The return of information is particularly important as it allows the application program to make a decision as to whether to wait for the resource or take some other action.

The problem of preventing lower priority tasks locking out higher priority tasks through the monitor access mechanism has been tackled in a number of ways. One solution adopted in some implementations of Modula has been to run a monitor with all interrupts locked out; hence a monitor function, once invoked, runs to completion. In many applications, however, this can be too restrictive and some implementations allow the programmer to set a priority level on a monitor such that all lower priority tasks are locked out – it should be noted that this is an interrupt priority level not a task priority.

The monitor has proved to be a popular idea and in practice it provides a good solution to many of the problems of concurrent programming. The benefits and popularity of the monitor construct stems from its modularity which means that it can be built and tested separately from other parts of the system, in particular, from the tasks which will use it. Once a fully tested monitor is introduced into the system the integrity of the data or resource which it protects is guaranteed and a fault in a task using the monitor cannot corrupt the monitor or the resource which it protects. Although it does rely on the use of signals for inter-task synchronization it does have the benefit that the signal operations are hidden within the monitor.

A problem which restricts the usefulness of the monitor construct in some real-time applications is that a task leaving a monitor can only signal and awaken one other task, otherwise the requirement that only one task be active within a monitor will be contravened. One such application would be a system with a single controlling synchronizer task, e.g., a clock-level scheduler built as a monitor. The problem can be avoided by allowing signals to be used outside a monitor but then all the problems associated with signals and semaphores reemerge.

A second problem is that the monitor is an ideal vehicle for creating abstract mechanisms which fit in well with the idea of top-down design. The use of monitors in the top-down approach, however, leads to a requirement to call procedures in one monitor from within another monitor. This leads to the nested monitor call problem which can lead to deadlock. Provided nested monitor calls are prohibited the use of the monitor concept provides a satisfactory solution to many of the problems for a single processor machine or for a multi-processor machine with shared memory. It can also be used on distributed systems.

7.6 RENDEZVOUS

As an alternative to the use of monitors and signals to ensure mutual exclusion and synchronization in inter-task communication the idea of a rendezvous has been developed [Hoare 1978 and Brinch Hansen 1978]. In the rendezvous the actions of synchronization and data transmission are seen as inseparable activities. The fundamental idea is that if two tasks A and B wish to exchange data, e.g., if A wishes to transmit data to B, then A must issue a transmit request and B a receive request. If Task A issues the transmit request before B has issued the receive then A must wait until B issues its request and vice versa. When both tasks have synchronized the data is transferred and the tasks can then proceed independently.

The problem which arises with the original formulation is that both tasks must name each other and hence it is not possible to create general library tasks. The solution adopted in the language Ada is to use 'asymmetric' rendezvous in which only one task, known as the 'caller', names the other task, known as the 'server'. In the descriptions that follow it is assumed that the language which supports the rendezvous concept does so by means of a construct which takes the form

```
ACCEPT name(parameter list)
   ...
   statements
   ...
END
```

The statements within the ACCEPT .. END are assumed to be a critical section and are executed in a mutually exclusive manner. They would normally be executed by the server task. The ACCEPT statement represents an entry point and the calling task specifies the name of the entry point when it wishes to synchronize with the server task.

* * *

Example 7.14
```
TASK A;
  VAR x:ADataItem;
    BEGIN
      ...
      B.Transfer(x);
      ...
    END;
TASK B;
  VAR y:ADataItem;
    BEGIN
      ...
      ACCEPT Transfer(IN item:ADataItem);
        y:=item;
      END;
    END;
```

* * *

In the above example Task A wishes to pass information held in variable x to a variable y in Task B. The actual data transfer takes place using the normal parameter passing mechanisms: the actual parameters supplied in the call, in this case the variable x, are bound to the formal parameters of the ACCEPT statement, in this case *item*. The synchronization of the two tasks is obtained by the requirement that the entry procedure call – B.Transfer (x) – cannot be completed until the corresponding ACCEPT statement – ACCEPT Transfer – is executed and conversely the execution of the ACCEPT statement cannot be completed until the entry call is executed. The actual transfer is completed within the body of the ACCEPT statement; in this case the data supplied by the entry call is transferred to a variable which is local to Task B.

Note that in the ACCEPT statement the direction of the transfer is specified, in this case IN. Variables can be declared as being for input (IN) or output (OUT) or as bidirectional (IN OUT).

A problem which arises with the rendezvous concept is that the two tasks have to synchronize in order to transfer information. The task which is producing the information cannot leave it in a buffer and continue: it must wait for the consumer task to arrive before the transfer can take place. The position is equivalent to that which the motorist would face if there were no filling stations, only roving petrol tankers. The motorist and the petrol tanker driver would have to arrange for a rendezvous; when both arrived at the designated place the motorist would fill up with petrol from the tanker, then both would continue on their respective ways. The respective tasks would be as follows.

* * *

Example 7.15

```
TASK Car;
   VAR     x:AFuel;
   BEGIN
     ...
     Tanker.get(x);
     ...
   END;

TASK Tanker;
   VAR     y:AFuel;
   BEGIN
     ...
     ACCEPT get(OUT fuel:AFuel);
         fuel:= y;
     END;
   END;
```

* * *

Since in practice neither the motorist nor the tanker driver would wish to be restricted to having to meet at a particular time or place then this is not a satisfactory solution. Similarly the requirement of the rendezvous that tasks synchronize in order to exchange data is too severe a constraint for many applications.

The solution to strict synchronization is to introduce a buffer task between the two tasks which wish to exchange data. Continuing the example of the motorist and the petrol tanker, the buffer task will be AFillingStation. The task AFillingStation has two entry points, 'receive' and 'deliver'. The tanker calls the entry point 'deliver' in order to transfer petrol from the tanker to the filling station tanks and the motorist calls 'receive' in order to obtain fuel from the filling station pumps.

* * *

Example 7.16

```
TASK Car;                         TASK Tanker;
   VAR    x:AFuel;                   VAR     y:AFuel;
   BEGIN                             BEGIN
     ...                               ...
     AFillingStation.receive(x)       AFillingStation.deliver(x);
     ...                               ...
   END;                              END;

TASK AFillingStation;
  VAR   y:AFuel;
  BEGIN
    LOOP
       ACCEPT deliver(IN fuel:AFuel);
           y:=fuel;
       END;
```

```
      ACCEPT receive(OUT fuel:AFuel);
        fuel:=y;
      END;
    END LOOP;
END;
```

* * *

In the above example the tanker makes its rendezvous with the filling station and delivers petrol into the tanks of the filling station and similarly the motorist rendezvous with the filling station and gets petrol from it. The problem with the task AFillingStation as given above is that the actions of deliver and receive have to be performed alternately which is of course impractical since the tanker will deliver a much larger quantity of fuel than a single motorist will receive. The problem is solved by introducing non-determinism into the task AFillingStation.

* * *

Example 7.17

```
TASK AFillingStation;
  VAR   y:AFuel;
  BEGIN
    DO
      SELECT
        ACCEPT deliver(IN fuel:AFuel);
          y:=fuel;
        END;
      OR
        ACCEPT receive(OUT fuel:AFuel);
          fuel:=y;
        END;
      END SELECT;
    END DO;
  END;
```

* * *

The key to the operation of the above task is the action of the select statement: each time the select statement is executed there are four possible states in which the entry points to the task can be:

1. a call to deliver is pending;
2. a call to receive is pending;
3. calls to deliver and to receive are pending; or
4. no calls are pending.

For cases (1) and (2) the appropriate ACCEPT statement is immediately executed. In case (3) one of the ACCEPT statements is selected at random and executed; in case (4) the task is suspended until a call is made to either of the ACCEPT statements, at which time the task is resumed and the appropriate statement is executed. The select statement ensures that only one ACCEPT statement will be executed at any one time but the order is not predetermined and there can be successive calls to the same ACCEPT.

As was seen in Example 7.6, involving the transfer of control parameters from an input task to the control task itself, there is a need to be able to test if another task is waiting, or if another task has left data to be collected in order to avoid committing a high priority repetitive task to wait for an event. There is also frequently the requirement in real-time systems to have some form of time-out such that a task only commits itself to wait for a predetermined length of time. Two extensions to the rendezvous primitive provide facilities to support these actions.

The time-out facility is provided in a simple and natural way by extending the select statement to allow a delay option in the possible choices within the select construct. This is illustrated below for the control parameter problem. It is assumed that when an input task has gathered the new parameters it makes a call to a put entry point in the control task. The control task includes the following code.

* * *

Example 7.18

```
TASK Control;
    ...
BEGIN
    ...
    (*control action*)
    ...
  (*start of section to check if update of parameters is required*)
    SELECT
        ACCEPT Put(IN parameters:AControlParRecord);
            kp := parameter.kp;
            kd := parameter.kd;
            ki := parameter.ki;
        END
    OR
        DELAY 1    (*delay in milliseconds*)
        ...
        END
    END
END.
```

* * *

In the above code fragment if, when the control task reaches the select statement a call to the entry point Put is pending, then the accept part of the select statement is executed and the parameter values are transferred to the control task. However, if no call is pending then the delay part of the select statement is executed. The action of the delay is to cause the control task to wait for the length of time specified in the delay; during this period of suspension any call to the accept statement will be recognized and the accept statement executed. If no calls are received, then at the end of the delay period the statements following the delay statement are executed.

An alternative to the delay part within a select statement is an else part (this can be thought of as a delay 0). The above problem could be coded using the else statement as follows.

* * *

Example 7.19

```
TASK Control;
    ...
BEGIN
    ...
    (*control action*)
    ...
    (*start of section to check if update of parameters is required*)
    SELECT
            ACCEPT Put(IN parameters:AControlParRecord);
                kp := parameter.kp;
                kd := parameter.kd;
                ki := parameter.ki;
            END
        OR
            ELSE
            ...
            END
    END
END.
```

* * *

In Example 7.19, if there is no call pending for the accept statement when the select statement is reached, then the else part of the select statement is executed immediately. The use of the SELECT .. OR .. ELSE construct is the most appropriate for the control parameters problem.

The delay statement is useful in many applications, e.g., on detection of an alarm condition the required action may be to alert the operator and to expect the operator to acknowledge the presence of the alarm and take appropriate action within a predetermined time. If the operator does not respond then the computer system has to take further action, possibly by sounding an audible alarm or by beginning to close down the plant. The SELECT .. DELAY construct provides a natural and simple way of expressing the requirement.

* * *

Example 7.20

```
SELECT
        ACCEPT OperatorAcknowledge;
    OR
        DELAY 30 (*delay 30 seconds*)
        ...
        AlternativeAction;
        ...
END
```

* * *

Another example of the use of the delay statement is to provide time-out in communications with peripherals or other computers.

The rendezvous concept provides the most flexible and easily understood mechanism for handling multi-tasking problems and in the select mechanism provides facilities which none of the other concepts have. It has been implemented as part of the Ada language.

7.7 SUMMARY

The study of concurrent programming has clarified many of the problems which have had to be solved in the implementation of real-time control systems and has led to the development of high-level languages which support the building of multi-tasking software. The techniques described above have been applied to the implementation of real-time systems in two main ways.

1. The addition of support for semaphores and signals to existing languages – typically to Pascal – with the provision of a small real-time executive and a range of operating system modules. A typical example of this approach is the Texas Instrument Pascal system.
2. The development of languages which are suitable for the construction of operating systems as well as for application programs, e.g., Modula-2.

The work has also provided the means by which modular operating systems can be created. The majority of such systems have been built to operate on top of existing operating systems rather than on a bare machine. The outline of part of a modular operating system, based on the use of Modula-2, is described in Chapter 9.

EXERCISES

7.1 In the heat process example discussed in Chapter 5 provision was made for normal output to the display and for alarm output. If the alarm output is restricted to a specific area of the screen does the display need to be allocated either to the standard task or to the alarm task in a mutually exclusive manner? If the device is not allocated mutually exclusively what precautions will be required in the driver software to prevent a mix up of information on the display?

7.2 Information on the orders for widgets received by a manufacturing company is held on a disk file. Each record on the file corresponds to an order and contains the following:

(a) code for customer [integer 0 .. maxint];
(b) size of widget [integer 0 .. 100];
(c) grade [A .. Z];
(d) number ordered [integer 0 .. maxint]; and
(e) date of order [dd:mm:yyyy].

A program has to be provided that enables the manger to obtain for a period between any two dates any of the following:

(i) total number of widgets of a particular size;
(ii) total number of widgets of each grade ordered;
(iii) distribution of grades; and
(iv) distribution of sizes.

Show in outline the design of the software to obtain the information. How would the design have to be changed if the operations (i), (ii), (iii) and (iv) were able to be performed concurrently for different ranges of dates? Is there any advantage in performing the operations concurrently on a single processor?

7.3 For the system described in Exercise 7.2, additional terminals are to be provided to enable the sales manager, general manager and sales director to obtain the information online. Design a suitable program (in outline).

REFERENCES AND BIBLIOGRAPHY

ALLWORTH, S.T. (1981), *Introduction to Real-time Software Design*, Macmillan

BEN-ARI, M. (1982), *Principles of Concurrent Programming*, Prentice Hall

BENNETT, S. and LINKENS, D.A. (eds.) (1984), *Real-time Computer Control*, Peter Peregrinus, Stevenage

BUDGEN, D. (1985), 'Combining MASCOT with Modula-2 to aid the engineering of real-time systems', *Software: Practice and Experience*, **15(8)**, pp. 767–93

CASSELL, D.A. (1983), *Microcomputers and Modern Control Engineering*, Reston Publishing Co

DAHL, O.J., DIJKSTRA, E.W. and HOARE, C.A.R. (1972), *Structured Programming*, Academic Press

DIJKSTRA, E.W. (1968), 'Cooperating sequential processes', *Programming Languages*, Academic Press

HENRY, R. et al. (1984), *The ModOS User's Manual*, Human-Computer Interaction Group, Psychology Department, Nottingham University

HOLT, R.C., GRAHAM, G.S., LOZOWSKA, E.D. and SCOTT, M.A. (1978), *Structured Concurrent Programming with Operating Systems Applications*, Addison Wesley

LISTER, A.M. (1979), *Fundamentals of Operating Systems*, Macmillan (2nd Edition)

SEARS, K.H. and MIDDLEDITCH, A.E. (1985), 'Software concurrency in real-time control systems: a software nucleus', *Software: Practice and Experience*, **15(8)**, pp. 739–59

SIMPSON, H.R. and JACKSON, K. (1979), 'Process synchronization in MASCOT', *Computer Journal*, **22(4)**, pp. 332–45

WIRTH, N. (1977), 'Towards a discipline of real-time programming', *Comm. ACM*, **20(8)**, pp. 577–83

WIRTH, N. (1977), 'Modula: a language for multiprogramming', *Software: Practice and Experience*, **7(1)**, pp. 3–84

WIRTH, N. (1982), *Programming in Modula-2*, Springer

YOUNG, S.J. (1982), *Real-time Languages: Design and Development*, Ellis Horwood

YOUNG, S.J. (1983), *An Introduction to Ada*, Ellis Horwood

8

Real-time Languages

8.1 INTRODUCTION

The requirements of real-time software place heavy demands on programming languages. It should by now be obvious that it is essential that real-time software is reliable: the failure of a real-time system can be expensive both in terms of lost production, or in some cases, in the loss of human life (e.g., through the failure of an aircraft control system). Real-time systems are frequently large and complex, a factor which makes development and maintenance costly. Such systems have to respond to external events with a guaranteed response time; they also involve a wide range of interface devices, including non-standard devices. In many applications efficiency in the use of the computer hardware is vital in order to obtain the necessary speed of operation.

The demands placed on a language used for the implementation of real-time software are severe and they are discussed in detail in the next section. Traditionally assembly-level languages have been used for the construction of real-time software, largely because of the need to obtain efficient use of the CPU and to provide the ability to access interface devices and support interrupts. Assembly coding is still widely used for small systems with very high computing speed requirements, or for small systems which will be used in large numbers. In the latter case the high cost of development is offset by the reduction in unit cost through having a small, efficient, program.

The requirements of a real-time language subsume the requirements of a general purpose language and so many of the features described below are also present (or desirable) in languages which do not support real-time operations. The starting point for determining the features required in a language is to consider the *user requirements*; these can then be converted into the *language requirements* needed to meet the user requirements, and finally the features required to meet the language requirements can be determined.

8.2 USER REQUIREMENTS

This section follows the discussion given by Barnes [1976] and Young [1982] who divided the user requirements into six general areas:

User Requirements

1. security;
2. readability;
3. flexibility;
4. simplicity;
5. portability; and
6. efficiency.

8.2.1 Security

Security can be considered to be a measure of the extent to which a language is able to detect errors automatically either at compile time or through the run-time support system. It is rarely possible to test software exhaustively (for those who doubt this statement Pressman [1982], pages 292–3 provides a simple supporting example) and yet a fundamental requirement of real-time systems is that they operate reliably. The intrinsic security of a language is therefore of major importance for the production of reliable programs.

Economically it is important to detect errors at the compilation stage rather than at run-time since the earlier the error is detected the less it costs to correct it. In real-time system development the compilation is often performed on a different computer than the one used in the actual system, whereas run-time testing has to be done on the actual hardware and, in the later stages, on the hardware connected to plant. Run-time testing is therefore expensive and can interfere with the hardware development program. Reliance on run-time checking frequently requires that additional code is inserted in the program (normally done by the compiler) and this leads to an increase in program size and a decrease in execution speed.

A typical example of the need for run-time checking is the checking of array bounds. Many compilers allow the insertion of check code as an option which is selected when the program is compiled. This is an acceptable approach for standard programming: during the initial testing the array bound check is selected but once the program appears to function correctly it is omitted and hence the program runs faster. Although the facility is often used in this way for real-time systems it is not a reliable solution. (Why not?)

There are two ways to do the array checking. Consider an array of 20 numbers denoted by array[i] where i is an integer variable. For security the system must ensure that whenever array[i] is used the condition $1 <= i <= 20$ is satisfied. The first way of checking is simply to insert a check before every array access to ensure that the condition is not violated. An alternative technique used by the more modern compilers is to check the condition whenever an assignment is made to i. The use of the second method requires the co-operation of the programmer who must specify the permitted range of variable i when it is declared. Provided that the checking of the assignment of values to i is done then there is no need to check access to array[i]. A further reduction in run-time checking is also found with this technique in that a large number of the assignments to i can be tested during compilation. The second method relies on the language being strongly typed (this is explained below). In general, strong typing leads to high compile-time security.

It should be stressed that there are some errors which cannot be detected by any language: for example, errors in the logical design of the program. The chance of such errors occurring is lessened if the language encourages the programmer to write clear, well-structured, code. Hence a further requirement for security is that the language must be clear, readable and well-structured.

8.2.2 Readability

The readability of a program is a measure of the ease with which its operation can be understood without resort to supplementary documentation such as flowcharts or natural language descriptions. The emphasis is on ease of reading because a particular segment of code will only be written once but will be read many times. The benefits of good readability are:

- Reduction in documentation costs: the code itself provides the bulk of the documentation. This is particularly valuable in projects with a long life expectancy in which inevitably there will be a series of modifications. Obtaining up-to-date documentation and keeping documentation up-to-date can be very difficult and costly.
- Easy error detection: clear readable code makes errors, e.g., logical errors, easy to detect and hence increases reliability.
- Easy maintenance: it is frequently the case than when modifications to a program are required the person responsible for making the modifications was not involved in the original design – changes can only be made quickly and safely if the operation of the program is clear.

The factors which affect readability are manifold and to some extent readability depends on personal preference. The cooperation of the programmer is also required: it is possible to write unreadable programs in any language. The readability of a program is improved by the adoption of a clear layout which emphasizes the structure, and by the careful choice of variable names. The facility to use upper and lower case characters and long variable names enhances readability; e.g.

outputTemp := 75

is much clearer and understandable than

OUTTMP = 75

An example of a BASIC program, in which close attention to layout has been paid, is given in Figure 8.1. The program shows the code required to provide three-term control of a simple process. It is assumed that an 8-bit analog-to-digital converter located at memory address 04H is used to measure the output of the process and that an 8-bit digital-to-analog converter located at address 06H is used for the signal controlling the process. A timing signal is provided that sets bit 0 of a

```
10 REM PID CONTROL
20 REM ************************************************************
30 REM
40 REM   S BENNETT    21 JULY 1986
50 REM
60 REM ************************************************************
70 REM
80 A$="BEX8A" : REM FILE NAME
90 REM
100 REM DEFINE VARIABLE USUAGE
110 REM
120    A =255  : REM ADC RANGE
130    B =255  : REM DAC RANGE
140    C =4    : REM ADC CHANNEL
150    D =0.1  : REM DERIVATIVE ACT
160    E =0    : REM ERROR VALUE
170    E1=0    : REM OLD VALUE OF ERROR
180    F =6    : REM DAC ADDRESS
190    G =2    : REM TIMER ADDRESS
200    I =0.1  : REM INTEGRAL GAIN
210    M =0    : REM MEASURED VALUE
220    M1=100  : REM MEASURED VARIABLE RANGE
230    P =0.2  : REM PROPORTIONAL GAIN
240    R =50   : REM SETPOINT VALUE
250    S =0    : REM SUMMATION VALUE
260    S1=150  : REM MAXIMUM VALUE OF S
270    S2=-50  : REM MINIUMUM VALUE OF S
280    U =50   : REM CONTROL VALUE
290    U1=255  : REM MAXIMUM CONTROL VALUE
300    U2=0    : REM MINIMUM CONTROL VALUE
310    U3=50   : REM STEADY STATE OFFSET
320    U4=100  : REM CONTROL OUTPUT RANGE
330 REM
340 REM MAIN LOOP ********************************************
350 REM
360        M=PEEK(C)
370        M=M*M1/A
380        E=R-M
390        S=S+E
400        IF S>S1 THEN S=S1    : REM INTEGRAL WIND-UP CHECK
410        IF S<S2 THEN S=S2
420        U=P*E + I*S + D*(E-E1) + U3
430        U=U*B/U4
440        IF U>U1 THEN U=U1    : REM OUTPUT LIMIT CHECK
450        IF U<U2 THEN U=U2
460        POKE(F,U)
470        E1=E
480 REM
490 REM  WAIT FOR NEXT SAMPLE INTERVAL
500 REM
510        T=PEEK(G)
520        IF T=0 THEN GOTO 510
530        T=255
540        POKE(G,T)
550        GOTO 360
560 REM
570 REM END OF MAIN LOOP *************************************
580 REM
590 END
```

Fig. 8.1 Example of BASIC program – clear layout.

register located at address 02H every 0.5 seconds. The bit is reset by writing the value FFH to location 02H. The range 0..255 of the analog-to-digital converter corresponds to 0..100% of the input signal and similarly the range of the digital-to-analog converter 0..255 corresponds to 0..100% of the control signal.

Figure 8.2 shows the identical program written with no attention to readability. In Figure 8.3, a similar program written for the BBC Microprocessor is listed: it shows the advantage, in terms of readability, of long names and the ability to mix upper and lower case characters.

```
10REM PID CONTROL
20A$="EX8B"
30A=255:B=25:C=4:D=0.1:E=0:E1=0:F=6:G=2:I=0.1:M=0:M1=100:P=0.2
40R=50:S=0:S1=150:S2=-50:U=50:U1=255:U2=0:U3=50:U4=100
50M=PEEK(C):M=M*M1/A:E=R-M:S=S+E:IF S>S1 THEN S=S1
60IFS<S2THENS=S2
70U=P*E+I*S+D*(E-E1)+U3:U=U*B/U4:IF U>U1 THEN U=U1
80IFU<U2THENU=U2
90E1=E:POKE(F,U)
100T=PEEK(G):IFT=0THEN100
110T=255:POKE(G,1)
120END
```

Fig. 8.2 Example of BASIC program – poor layout.

Languages with clear, well understood, control structures such as IF..THEN ..ELSE and REPEAT..UNTIL aid readability. The provision of higher level facilities supporting modularization and control of object visibility also helps to improve readability. In particular, it is easier to understand a program if it is divided into self-contained operational units which can be understood without reference to other parts of the program (examples of division into modules are given in Chapter 9). A major difficulty which arises in program maintenance is through changes in one section causing errors in another section which has not been altered.

The readability of three different languages (assembly, Pascal and C) can be assessed from Figures 8.5, 8.6 and 8.7, the flow chart for the program is given in Figure 8.4.

8.2.3 Flexibility

For a language to be described as a general purpose language there is a requirement that the programmer should be able to express all the operations required in a program without the need to use assembly coding. The flexibility of a language is a measure of this facility. It is particularly important in real-time systems, in that frequently non-standard I/O devices will have to be controlled. The achievement of high flexibility can conflict with achieving high security. The compromise that is reached in the modern languages is to provide high flexibility but through the module or package concept provide a means by which the low-level

```
 10 REM PID CONTROL
 20 REM ***********************************************************
 30 REM
 40 REM   S BENNETT   21 JULY 1986
 50 REM
 60 REM ***********************************************************
 80 A$="BEX8C"
 90 REM DEFINE VARIABLE USUAGE
100     ADCChan=1          : REM ADC CHANNEL ADDRESS
110     ADCRange=255       : REM ADC RANGE
120     controlRange=100   : REM CONTROL OUTPUT RANGE
130     DACAddress=&FE60   : REM ADDRESS OF DAC
140     DACRange=255       : REM DAC RANGE
150     e=0                : REM ERROR VALUE
160     e1=0               : REM OLD VALUE OF ERROR
170     kd=0.1             : REM DERIVATIVE ACTION
180     ki=0.1             : REM INTEGRAL GAIN
190     kp= 0.2            : REM PROPORTIONAL GAIN
200     m =0               : REM MEASURED VALUE
210     mRange=100         : REM MEASURED VARIABLE RANGE
220     nextTime%=0        : REM NEXT RUN TIME FOR CONTROL
230     now%=0             : REM CURRENT TIME
240     offset=50          : REM STEADY STATE OFFSET
250     r =50              : REM SETPOINT VALUE
260     sampleInt%=30      : REM SAMPLING INTERVAL IN CENTI-SECONDS
270     sum =0             : REM SUMMATION VALUE
280     sumMax=150         : REM MAXIMUM VALUE OF S
290     sumMin=-50         : REM MINIUMUM VALUE OF S
300     u=50               : REM CONTROL VALUE
310     uMax=255           : REM MAXIMUM CONTROL VALUE
320     uMin=0             : REM MINIMUM CONTROL VALUE
330 REM
340 REM ***********************************************************
350 REM
360 REM MAIN LOOP
370 REM
380     now%=TIME
390     m=ADVAL(ADCChan)
400     m=m*mRange/ADCRange
410     e=r-m
420     sum=sum+e
430     IF sum>sumMax THEN sum=sumMax
440     IF sum<sumMin THEN sum=sumMin
450     u=kp*e + ki*sum + kd*e1
460     u=u*DACRange/controlRange
470     IF u>uMax THEN u=uMax
480     IF u<uMin THEN u=uMin
490     ?DACAddress=u
500     e1=e
510 REM
520 REM    DELAY LOOP TO WAIT NEXT SAMPLE TIME
530 REM
540     nextTime%=now%+sampleInt%
550     IF nextTime%<TIME THEN PROCTimeError:STOP
560     REPEAT UNTIL TIME>=nextTime%
570     GOTO 380
580 REM
590 REM END OF MAIN LOOP
600 REM
610  DEF PROCTimeError
620     PRINT "TIMING ERROR   sampleInterval=";sampleInt%
630  ENDPROC
>
```

Fig. 8.3 BASIC program of Figure 8.1 rewritten for BBC BASIC.

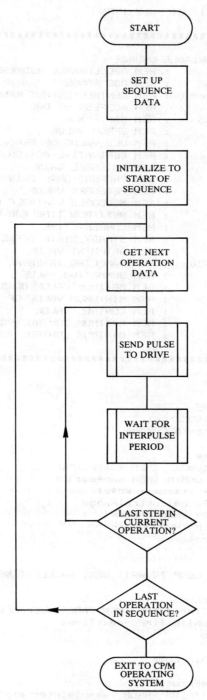

Fig. 8.4 Flow chart for stepper motor drive (reproduced from Hoyle, *Electronics and Power*, **30**, 1984).

```
        .processor       z80          ;for 'universal' cross-assembler
;
; Stepper motor drive program
;
        .psect  _data                 ;start of data area
;
; Constant data:-
;
        .define drive    = 18h,       ;i/o port number
                acw      = 2,         ;direction bit (1) set
                cw       = 0,         ;direction bit (1) reset
                exit     = 0          ;CP/M re-start address
;
; Variable data record (all limited to 1 byte for brevity)
;
;    1st operation:-
ops:            .byte    [1]          ;number of steps (0..255)
                .byte    [1]          ;direction 0:cw, 1:acw
                .byte    [1]          ;inter-pulse delay (0..255)
;    2nd operation (etc. etc....):-
;                |
;    Ancillary data :-
numops:         .byte    [1]          ;total number of operations
dirn:           .byte    [1],0        ;current direction
;
        .psect  _text                 ;start of code area
;
; Set up sequence of operations - omitted
;                |
;
;·Main loop for specified number of operations
;
                ld       b,numops     ;operations count
                ld       hl,ops       ;start of record
oloop:  push    bc                    ;save count
                ld       b,(hl)       ;get no_of_steps
                inc      hl
                ld       a,(hl)       ;get direction
                inc      hl
                ld       c,(hl)       ;get delay
                inc      hl
                sub      acw          ;test direction in a
                jr       nz,scw       ;go to set clockwise
                ld       a,acw        ;set to anticlockwise
                jr       setd
scw:            ld       a,cw
setd:           ld       (dirn),a     ;save set direction
;
; Secondary loop for specified number of steps
;
sloop:  call    pulse                 ;send a pulse to the drive
        call    wait                  ;inter-pulse period
; test for last step
                djnz     sloop        ;repeat if not
; test for last operation
                pop      bc           ;restore count
                djnz     oloop        ;repeat if more operations
                jp       exit         ;to external operating system
;
; Subroutine to pulse stepper drive
;
pulse:  ld      a,(dirn)              ;get direction
        set     0,a                   ;set step bit
        out     drive,a
        res     0,a                   ;reset step bit
        out     drive,a
        ret
.end
```

Fig. 8.5 Assembly level coding for stepper motor drive (reproduced from Hoyle, *Electronics and Power*, **30**, 1984).

```
PROGRAM stepper_drive:

{ Define data structure }

CONST
    max_operations = 10;   { arbitrary limit to sequence }
TYPE
    operation =
        RECORD
            number_of_steps : integer;
            direction : boolean; { true or false }
            delay: integer { inter-pulse }
        END; { of record specification }
VAR
    sequence : ARRAY[1..max_operations] OF operation;
    number_of_operations : integer; { actually in sequence }
    this_step, this_operation : integer;

PROCEDURE pulse(direction : boolean);    { stepper drive routine }

    CONST
        { define drive interface bits and direction values }
        dirn_bit = 1;           clockwise      = FALSE;
        step_bit = 0;           anticlockwise = TRUE;
    VAR
        drive : integer;    { control output word }

    { External library routines - bit set/reset, digital output }
    FUNCTION pfset(bit,wordofbits:integer):integer;EXTERNAL;
    FUNCTION pfres(bit,wordofbits:integer):integer;EXTERNAL;
    PROCEDURE ppdout(outword:integer);EXTERNAL;

    BEGIN
        drive := 0;
        IF direction = clockwise
            THEN drive := pfres(dirn_bit,drive)
            ELSE drive := pfset(dirn_bit,drive);
        drive := pfset(step_bit,drive);      { pulse hi }
        ppdout(digout);
        drive := pfres(step_bit,drive);      { pulse lo }
        ppdout(drive)
    END; { of procedure pulse }

BEGIN { main program }

    { Set up sequence of operations - omitted }
    {                |                        }

    { Execute sequence of operations one by one }
    FOR this_operation := 1 TO number_of_operations DO
      WITH sequence[this_operation] DO
        BEGIN
            FOR this_step := 1 TO number_of_steps DO
                BEGIN
                    pulse(direction);
                    wait(delay)
                END

        END
END.
```

Fig. 8.6 Stepper drive control in Pascal (reproduced from Hoyle, *Electronics and Power*, **30**, 1984).

```c
/*
 *  Stepper motor drive program
 *
 *
 *  Constant data and record 'template' :-
 *
 */
#define MAX_OPERATIONS  10           /* arbitrary limit  */
#define SETPULSE        1            /* interface masks  */
#define RESETPULSE      !(SETPULSE)
#define ACW_SET         2
#define CW_SET          0
#define DRIVEPORT       0x18         /* i/o port address */

struct operation
{
    int  steps;
    char dirn;         /* cw: 0, a/cw: 1 */
    int  delay;        /* inter-pulse    */
};

main()
{
    /*
     * Variable data and main program :-
     */
    struct operation seqn[MAX_OPERATIONS];
    struct operation *this_op;         /* to point to seqn[.]  */
    int no_of_ops;                     /* actually in sequence */
    int this_step;

    /* Set up sequence of operations (seqn) - omitted    */
    /*                                                   */

    /* execute operations one by one */
    for (this_op = seqn; this_op != (seqn + no_of_ops); this_op++)
    {
        for (this_step = 0; this_step < (this_op->steps); this_step++)
        {
            pulse(this_op->dirn);
            wait(this_op->delay);
        }
    }
}

pulse(direction)      /* stepper drive routine */
    char direction;
{
    int drive;
    extern out();     /* external library routine */

    drive = 0;
    if (direction)
        drive |= ACW_SET;
    else drive &= CW_SET;
    drive |= SETPULSE;
    out(DRIVEPORT,drive);
    drive &= RESETPULSE;
    out(DRIVEPORT,drive);
}
```

Fig. 8.7 Stepper drive control in C (reproduced from Hoyle, *Electronics and Power*, **30**, 1984).

(i.e., unsecure) operations can be hidden in a limited number of self-contained sections of the program.

8.2.4 Simplicity

In language design, as in other areas of design, the simple is to be preferred to the complex. Simplicity contributes to security. It reduces the cost of training, reduces the probability of programming errors arising from misinterpretation of the language features, it reduces compiler size and leads to more efficient object code.

Associated with simplicity is consistency: a good language should not impose arbitrary restrictions (or relaxations) on the use of any feature of the language.

8.2.5 Portability

The achievement of portability, while very desirable as a means of speeding up developments, reducing costs, and increasing security, is difficult to achieve in practice. Surface portability has improved with the standardization agreements on many languages, i.e., it is now often possible to transfer a program from one computer to another and find that it will compile and run on the computer to which it has been transferred. There are, however, still problems when the word-lengths of the two machines differ: there may also be problems with the precision with which numbers are represented even on computers with the same word-length.

The problem of portability is frequently more difficult for real-time systems in that they often make use of specific features of the computer hardware and the operating system. A practical solution is to accept that a real-time system will not be directly portable, but to restrict the areas of non-portability to specific modules by restricting the use of low-level features to a restricted range of modules. Portability can be further enhanced by writing the application software to run on a virtual machine, rather than for a specific operating system. A change of computer and operating system then requires the provision of new support software to create a virtual machine on the new system.

8.2.6 Efficiency

In the early computer control systems great emphasis was placed on efficiency of the coding – both in terms of the size of the object code and in the speed of operation – as computers were both expensive and, by today's standards, very slow. As a consequence programming was carried out using assembly languages and frequently 'tricks' were used to keep the code small and fast. The desire for the generation of efficient object code was carried over into the designs of the early real-time languages and in these languages the emphasis was on efficiency rather than security and readability.

The falling costs of hardware and the increase in the computational speed of computers has changed the emphasis. Also in a large number of real-time applications the concept of an efficient language has changed to include considerations of the security and the costs of writing and maintaining the program: speed, and compactness of the object code have become of secondary importance. There are, however, still application areas where compactness and speed do matter; in the consumer market, where production runs may be 100 000 per year, the ability to use a slower, cheaper CPU or to keep down the amount of memory used can make a significant difference to the viability of the product.

Other areas in which speed matters are in control of electro-mechanical systems, aircraft controls and in the general area of signal processing, e.g., speech recognition.

Table 8.1 Comparison of size and execution speed of BASIC programs

Figure No.	File name	Size (bytes)	Time (ms)
8.1	BEX8A	1654	27.7
	EX8A	1488	27.7 (REM statements removed)
8.2	EX8B	350	26
8.3	EX8C	1803	28.8 (REM statements removed)
	BEX8C	2147	28.8
8.8	EX8F	1698	17 (Integer arithmetic)

One of the temptations when using an interpreted language such as BASIC is to sacrifice readability for efficiency. Table 8.1 shows the code size and execution times for a number of different implementations of a simple three-term control algorithm in BASIC. The code size and execution time for the readable programs (Figures 8.1 and 8.3) are larger and longer respectively than for the condensed version shown in Figure 8.2. A further gain in speed can be obtained if the program is rewritten to use only integer values. This is shown in the table and the program code is shown in Figure 8.8.

8.3 LANGUAGE REQUIREMENTS AND FEATURES

The ability to meet the user requirements outlined in the previous section easily is strongly dependent on the presence or absence of certain language features. The engineer faced with the choice of a programming language for a particular project needs to be aware of some of the technical details of different programming language features in order to make a rational decision. The major features which must be considered are listed below.

1. Declarations.
2. Types – including structured types and pointers.

```
>LIST
   10 REM PID CONTROL
   20 REM ************************************************************
   30 REM
   40 REM   S BENNETT   21 JULY 1986
   50 REM
   60 REM ************************************************************
   70 REM
   90 A$="EX8F"    : REM FILE NAME
  100     A% =255  : REM ADC RANGE
  110     B% =255  : REM DAC RANGE
  120     C% =4    : REM ADC CHANNEL ADDRESS
  130     D% =100  : REM DERIVATIVE ACTION
  140     E% =0    : REM ERROR VALUE
  150     E1%=0    : REM OLD VALUE OF ERROR
  160     I% =100  : REM INTEGRAL GAIN
  170     M% =0    : REM MEASURED VALUE
  180     M1%=100  : REM MEASURED VARIABLE R
  190     P% =200  : REM PROPORTIONAL GAIN
  200     R% =100  : REM SETPOINT VALUE
  210     S% =0    : REM SUMMATION VALUE
  220     S1%=255  : REM MAXIMUM VALUE OF S
  230     S2%=-255 : REM MINIUMUM VALUE OF S
  240     U% =50   : REM CONTROL VALUE
  250     U1%=255  : REM MAXIMUM CONTROL VALUE
  260     U2%=0    : REM MINIMUM CONTROL VALUE
  270     U3%=120  : REM STEADY STATE OFFSET
  290 REM
  300 REM MAIN LOOP ***********************************************
  310 REM
  320 X%=0:NOWTIME%=TIME
  330 REM     M=PEEK(C)
  340 X%=X%+1: IF X%>100THEN GOTO 510
  350       M%=120
  360       E%=R%-M%
  370       S%=S%+E%
  380       IF S%>S1% THEN S%=S1%
  390       IF S%<S2% THEN S%=S2%
  400       U%=P%*E% + I%*S% + D%*(E%-E1%) + U3%
  410       U%=(U%*B%) DIV 500000
  420       IF U%>U1% THEN U%=U1%
  430       IF U%<U2% THEN U%=U2%
  440 REM    POKE(B,U)
  450       E1%=E%
  460       GOTO 330
  470 REM
  480 REM END OF MAIN LOOP
  490 REM
  500 END
  510 ENDTIME%=TIME
  520 VDU 2
  530 PRINT A$
  540 PRINT "AVERAGE TIME(ms)="; (ENDTIME%-NOWTIME%)/10
  550 PRINT "PROGRAM SIZE="; (TOP-PAGE)
  560 VDU 3
  570 END
```

Fig. 8.8   Program of Figure 8.1 rewritten to use integer variables.

3. Initialization.
4. Constants.
5. Control structures.
6. Scope and visibility.
7. Modularity.
8. Exception handling.
9. Independent/separate compilation.
10. Multi-tasking.
11. Low-level constructs.

8.4 DECLARATIONS

The purpose of declaring an object used in a program is to provide the compiler with information on the storage requirements and to inform the system explicitly of the names being used. Languages such as Pascal, Modula-2 and Ada require all objects to be specifically declared and for a type to be associated with the object at declaration. The provision of type information allows the compiler to check that the object is used only in operations associated with that type. If, for example, an object is declared as being of type real and then is used as an operand in logical operation, the compiler should detect the type incompatibility and flag the statement as being incorrect.

Languages such as BASIC and FORTRAN do not require explicit type declarations: the first use of a name is deemed to be its declaration. In FORTRAN, explicit declaration is optional and objects can be associated with a type if declared; if objects are not declared then implicit typing takes place: names beginning with the letters I–N are assumed to be integer numbers, names beginning with any other letter are assumed to be real numbers.

Optional declarations are dangerous because they can lead to the construction of syntactically correct but functionally erroneous programs. Consider the following program fragment

```
100    ERROR=0
       ...
200    IF X=Y THEN GOTO 300
250    EROR=1
300    ...
       ...
400    IF ERROR=0 THEN GOTO 1000
       ...
```

In FORTRAN (or BASIC) ERROR and EROR will be considered as two different variables whereas the programmer's intention was that they should be the same: the variable EROR in line 250 has been mistyped. FORTRAN compilers cannot detect this type of error and it is a 'characteristic' error of FORTRAN. Many organizations which use FORTRAN extensively avoid such errors by insisting that all objects are declared and the code is processed by a pre-processor which checks that all names used are mentioned in declaration statements.

The implicit typing which takes place in FORTRAN can also lead to confusion and misinterpretation; consider the program fragment

```
REAL KP,KD,KI
INTEGER DACV,ADCV
100        DACV=KP*(KD/KI)
           END
```

A programmer reading the statement with the label 100 might assume that the operands of the statement are all integers and that the resultant value is then floated and assigned to a real variable, whereas the statement is doing just the opposite: the operands are real and the resultant is truncated and assigned to an integer variable.

8.5 TYPES

As we have seen above, the allocation of types is closely associated with the declaration of objects (in the above examples, variables). The allocation of a type defines the set of values that can be taken by an object of that type and the set of operations that can be performed on the object. The richness of types supported by a language and the degree of rigor with which type compatibility is enforced by the language are important influences on the security of programs written in the language. Languages which rigorously enforce type compatibility are said to be strongly typed: languages which do not enforce type compatibility are said to be weakly typed.

FORTRAN and BASIC are weakly typed languages: they enforce some type checking; e.g., the statements A$ = 25 or A = X$+Y are not allowed in BASIC, but they allow mixed integer and real arithmetic and provide implicit type changing in arithmetic statements. Both languages support only a limited number of types.

An example of a language which is strongly typed is Modula-2. In addition to enforcing type checking on standard types, Modula-2 also supports enumerated types. The enumerated type allows the programmer to define his or her own types in addition to using the pre-defined types. Consider a simple motor speed control system which has four settings: OFF, LOW, MEDIUM and HIGH and which is controlled from a computer system. Using Modula-2 the programmer could make the declarations:

```
TYPE AMotorState = (OFF,LOW,MEDIUM,HIGH);
VAR  motorSpeed : AMotorState;
```

The object motorSpeed can only have assigned to it one of the values enumerated in the TYPE definition statement; an attempt to assign any other value will be trapped by the compiler, e.g., the statement

```
motorSpeed := 150;
```

will be flagged as an error.

If we contrast this with the way in which the system could be programmed using FORTRAN we can see some of the protection which strong typing provides. In FORTRAN integers must be used to represent the four states of the motor control:

```
INTEGER OFF,LOW,MEDIUM,HIGH
DATA OFF/0/,LOW/1/,MEDIUM/2/,HIGH/3/
```

If the programmer is disciplined and uses only the defined integers to set MSPEED then the program is clear and readable, but there is no mechanism to prevent direct assignment of any value to MSPEED, hence the statements

```
MSPEED = 24
MSPEED = 150
```

would be considered as valid and would not be flagged as errors either by the compiler or the run-time system. The only way in which they could be detected is if the programmer inserted some code to check the range of values before sending them to the controller. In FORTRAN a programmer-inserted check would be necessary since the output of a value outside the range $0..3$ may have an unpredictable effect on the motor speed.

8.5.1 Sub-range types

Another valuable feature which enhances security is the ability to declare a sub-range of a type. In Modula-2 sub-ranges of ordinal types (i.e., INTEGER,CHAR and ENUMERATED types) can be defined. The following statements define sub-range types:

```
TYPE     ADACValue = 0..255;
    ALowerCaseChar = 'a'..'z';
```

and if the variables are defined as

```
VAR     output : ADACValue;
     character : ALowerCaseChar;
```

then the assignments

```
     output := -25;
  character := 'A';
```

will be flagged as errors by the compiler. The compiler will also insert run-time checks on all assignment statements involving sub-range types and any assignment which violates the permitted values will generate a run-time error. The use of sub-range types can increase the security of a program; however, in a real-time system full use of sub-range types may not be appropriate. They can be used if the run-time system permits transference of control to a user-supplied error analysis segment on detection of a run-time error; if it does not and it terminates execution of the program then the security of the system can be jeopardized by the use of sub-range types. Sub-range types can be useful during the development stages of the system as violations of correctly set sub-ranges can indicate logical errors in the code. For this reason many compilers provide an option switch to control the inclusion of sub-range checking.

Sub-range types have been extended in Ada to include sub-ranges of REAL. Again the usefulness is limited because of the extra code introduced in order to check for violations. In applications involving a large amount of computation the use of sub-ranges of REAL can slow down the computation significantly and hence in many applications the efficiency requirements will necessitate the use of explicit range checks at appropriate points rather than the use of compiler-supplied checks through the use of sub-range types.

8.5.2 Derived types

In many languages new types can be created from the implicit types: these are known as derived types and they inherit all the characteristics of the parent type. The use of derived types can make the meaning of the code clearer to the reader, for example:

```
TYPE          AVoltage = REAL;
              AResistance = REAL;
              ACurrent = REAL;
VAR           V1 : AVoltage;
              R3 : AResistance;
              I2 : ACurrent;
BEGIN
              I2 := V1/R3;
END;
```

In the above code the reader can easily see what is implied by the calculation, i.e., Ohm's law is being used to calculate the current flowing through a resistance. If the statement read

I2 := V1*R3;

then because I2 has been declared as of type ACurrent the reader would be suspicious that there was an error in the line. It should be noted that in languages such as Pascal and Modula the compiler would not detect errors of type incompatibility involving derived types (other than ones involving incompatibility of the parent types) since it would treat the derived types as identical to the parent types. This is not the case in Ada: typing is much stronger and although the derived types inherit the properties of the parent type they are treated as distinct types and hence a statement of the form

I2 := V1/R3;

would not be valid if I2, R3, and V1 had been declared as of types ACurrent, AResistance and AVoltage respectively. This does cause a problem and apparently limits the use of derived types since I2 = V1/R3 is a perfectly valid computation.

Ada, however, provides mechanisms for overcoming this problem and hence permitting the use of derived types.

8.5.3 Structured types

Most programming languages provide one structured type, the array. Arrays, though powerful, are limited in that all the elements of the array must be of the same type. There are many applications in which it would be useful to be able to declare objects which are made up of elements of different type.

* * *

Example 8.1

```
TYPE       AController = RECORD
                   inputAddress : INTEGER;
                  outputAddress : INTEGER;
                           gain : REAL;
                      maxOutput : REAL;
                           name : ARRAY [0..16] OF CHAR;
                         status : (OFF,ON);
                       END;
VAR        airFlow  : AController;
           fuelFlow : AController;
           reactant : AController;
PROCEDURE Control(VAR loop : AController) EXTERNAL;
BEGIN
    LOOP
        Control(airFlow);
        Control(fuelFlow);
        Control(reactant);
    END;
```

* * *

In the above example the whole of the information relevant to a particular control loop for the plant is contained in one object of type RECORD. The advantages of using a structure type such as the record structure is that the programmer does not continually have to consider the details of the way in which the information relevant to the object is stored: it contributes to the process known as abstraction.

In addition to RECORD many of the modern languages support types such as FILE and SET.

8.5.4 Pointers

Pointers are a mechanism by which objects can be referenced indirectly. They are widely used in systems programming and in some data processing applications. They can be used, e.g., to create linked lists and tree structures. Unrestricted availability of pointers, e.g., permitting pointers to objects of different types to be interchanged, can give rise to insecurity. Implicit in the use of pointers is the dynamic allocation and consequent de-allocation of storage: the overheads involved in the required mechanisms can be considerable and care in the use of pointer types is necessary in real-time applications.

8.6 INITIALIZATION

It is useful if, at the time of declaration of a variable, it can be given an initial value. This is not, of course, strictly necessary as a value can always be assigned to a variable. In terms of the security of a language it is important that the compiler checks that a variable is not used before it has had a value assigned to it. It is bad practice to rely on the compiler to initialize variables to some zero or null value. The security of languages such as Pascal and Modula-2 is enhanced by the compiler checking that all variables have been given an initial value.

8.7 CONSTANTS

Some of the objects referenced in a program will have constant values either because they are physical or mathematical entities such as the speed of light or pi; or because they are a parameter which is fixed for that particular implementation of the program, e.g., the number of control loops being used or the bus address of an input or output device. It is always possible to provide constants by means of initializing a variable to the appropriate quantity, but this has the disadvantage that it is insecure in that the compiler cannot detect if a further assignment is made which changes the value of the constant. It is also confusing to the reader since, unless the initial assignment is carefully documented, there is no indication which objects are constants and which are variables.

Pascal provides a mechanism for declaring constants, but since the constant declarations must precede the type declarations, only constants of the predefined types can be declared. This is a severe restriction on the constant mechanism in that for example it is not possible to do the following

A further restriction in the constant declaration mechanism in Pascal is that the value of the constant must be known at compilation time and expressions are not permitted in constant declarations. The restriction on the use of expressions in constant declarations is removed in Modula-2 (experienced assembler programmers will know the usefulness of being able to use expressions in constant declarations). For example in Modula-2 the following are valid constant declarations:

```
CONST
    message = 'a string of characters';
    length = 1.6;
    breadth = 0.5;
    area = length * breadth;
```

In Ada the value of the constant can be assigned at run-time and hence it is more appropriate to consider constants in Ada as special variables which become read-only following the first assignment to them.

8.8 CONTROL STRUCTURES

There has been extensive argument over the past few years about the use of both conditional and unconditional GOTO statements in high-level languages. It is argued that the use of GOTOs makes a program difficult to read and it has been shown that any program can be expressed without the use of GOTOs as long as the language supports the WHILE statement, the IF ... THEN ... ELSE conditional and BOOLEAN variables. Most modern languages support such statements. The standard structured programming constructs are shown in Figure 8.9.

From a theoretical point of view the avoidance of GOTOs is attractive; there are, however, some practical situations in which the judicious use of a GOTO can avoid complicated and confusing coding. An example of such a situation is when it is required to exit from a loop in order to avoid further processing when a particular condition occurs.

* * *

Example 8.2

A stream of data in character form is received from a remote station over a serial link. The data has to be processed character by character by a routine ProcessItem until the end-of-transmission character (EOT – ASCII code = 4) is received. The EOT character must not be processed.

A simple loop structure of the form

```
REPEAT
    get(character);
    ProcessItem(character);
UNTIL character = EOT;
```

cannot be used since the EOT character would be processed.
Possible solutions are:
  1.
```
    get(character);
    WHILE character <> EOT DO
        ProcessItem(character);
        get(character)
    END (* while *);
```

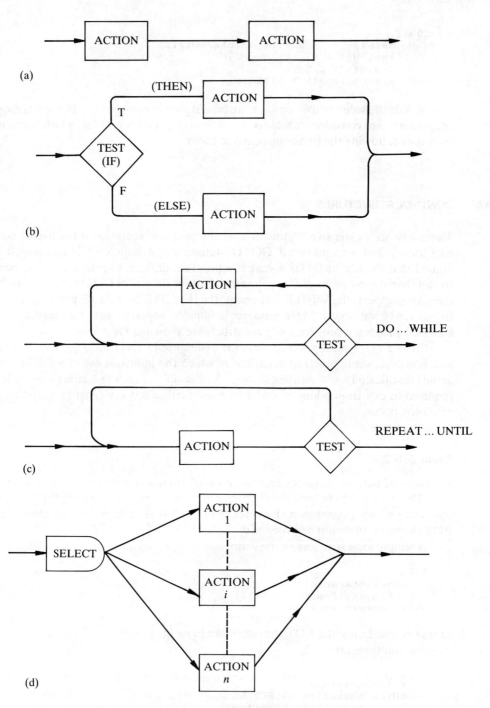

Fig. 8.9 Structured programming constructs. (a) Sequence; (b) decision; (c) repetition; (d) selection.

Scope and Visibility

2.
```
   finished := FALSE;
   REPEAT
     get(character);
     IF character = EOT THEN
         finished := TRUE
     ELSE
         ProcessItem(character)
     END (* IF *);
   UNTIL finished;
```

A much cleaner solution is provided by the use of an EXIT statement as in the fragment below

3.
```
   LOOP
      get(character);
      IF character = EOT THEN
          EXIT
      END (* if *);
      ProcessItem(character);
   END (* loop *);
(* EXIT causes a jump to statement here if EOT is detected *)
```

The solution shown in (3) which is possible in Modula-2 is much clearer because all the operations are shown within the loop statement whereas in solutions (1) and (2) some operation has to be performed outside the loop, i.e., either the first character has to be obtained before entering the loop or a Boolean variable has to be set before the loop is entered. The general LOOP ... END statement becomes particularly valuable if several different exceptions require an exit from the loop.

* * *

The need to be able to escape from a loop has been recognized in e.g. Modula-2 which provides a general non-terminating loop statement which can have in it one or more EXIT statements which cause control to pass to the statement immediately following the loop end. Similarly within procedures the statement RETURN causes an immediate exit from the procedure. A program illustrating the difference between RETURN and EXIT is shown in Figure 8.10.

8.9 SCOPE AND VISIBILITY

The scope of a variable is defined as the region of a program in which the variable is potentially accessible or modifiable. The regions in which it may actually be accessed or modified are the regions in which it is said to be visible. Thus in a FORTRAN program the scope of a variable declared in the main program extends over the whole program, but such a variable, unless named in a COMMON statement, will not be

```
MODULE ReturnExitLoop ;
(*
Title    : Example of different methods of leaving LOOP statement
File     : sb1 returne.mod
LastEdit : 24/3/87
Author   : S. Bennett
*)
FROM InOut IMPORT
     WriteString, WriteLn;
IMPORT Terminal;
VAR
        ch: CHAR;
    return : BOOLEAN;

PROCEDURE ReadKey ;
BEGIN
  WriteString('ReadKey called    ');
  LOOP
    WHILE NOT(Terminal.KeyPressed()) DO
      (* do nothing *)
    END (* while *);
    Terminal.Read(ch);
    IF ch='r' THEN
       RETURN
    END (* if *);
    IF ch='e' THEN
       EXIT
    END (* if *);
  END (* loop *);
  WriteString('this was EXIT not RETURN');
  WriteLn;
  return:=FALSE
END ReadKey;
BEGIN
  LOOP
    return:=TRUE;
    ReadKey;
    IF return THEN
       WriteString('this was RETURN not EXIT');
       WriteLn
    END (* if *);
    WriteString('To repeat type c, to stop any other character');
    WHILE NOT(Terminal.KeyPressed()) DO
      (* do nothing *)
    END (* while *);
    Terminal.Read(ch);
    IF ch<>'c' THEN
       EXIT
    END (* if *);
    WriteLn
  END (* LOOP *)
END ReturnExitLoop.
```

Fig. 8.10 Example showing behavior of RETURN and EXIT statements.

Scope and Visibility

visible in any sub-program. Scope and visibility are closely related to where in a program a variable is declared.

The argument over global or local declaration of objects has been almost as fierce as that over the use of GOTO statements. The proponents of local declarations argue that it is good practice to introduce names of objects close to where they are to be used and thus limit the scope of the object and its visibility. Those arguing in favor of global visibility of names claim that it is the only way in which consistency and control of the naming of objects can be achieved for large systems being developed by a team of programmers. They argue that local declaration leads to duplication of names and difficulties in subsequent maintenance of programs. A sensible compromise position is probably to declare globally the names of all objects which directly relate to the outside world, i.e., to the system being modeled or controlled, and to use local declaration for the names of all internal objects.

The way in which names of objects are declared is strongly influenced by the language being used. Most languages (with the exception of BASIC and COBOL) provide mechanisms for controlling scope and visibility. There are two general approaches: languages such as FORTRAN provide a single level of locality whereas the block-structured languages such as Pascal provide multi-level locality.

In FORTRAN the names of objects declared in a sub-routine are not visible outside that sub-routine. The main program can be considered as a parameterless sub-routine and names of objects declared in the main program are not visible in any of the sub-routines. This latter is a major difference from the block-structured languages in which the scope of objects declared at the outermost block (the main program) extends throughout all blocks. In FORTRAN global objects must be explicitly created by the use of a COMMON statement which must be repeated in every sub-routine which wishes to access such objects.

In the block-structured languages objects which are declared within a block may only be referenced inside that block. Blocks can be nested and the scope extends throughout any nested blocks. This is illustrated in Figure 8.11 which shows the scope for a nested PROCEDURE in Modula-2.

```
MODULE ScopeL1;
(*
Title    : Example of scope
File     : sb1 Scopel1.mod
LastEdit:
Author   : S. Bennett
*)
PROCEDURE L1;
VAR         A,B,C : INTEGER;
PROCEDURE L2;
    VAR         D,E,F : INTEGER;
    BEGIN
       (* variables A,B,C,D,E,F are visible here *)
    END  L2;
BEGIN
   (* variables A,B,C, visible here *)
END   L1;
(* no variables visible here *)
END ScopeL1.
```

Fig. 8.11   Scope of variables in nested procedures.

In a block-structured language an object declared in an inner block may have the same name as an object declared in an outer block. This does not cause any confusion to the compiler which simply provides new storage for the object in the inner block and the object in the outer block temporarily 'disappears', to reappear when the inner block is left. Although the compiler can easily handle reuse of names, it is not as easy for the programmer and the use of deeply nested PROCEDURE blocks with the reuse of names can compromise the security of a Pascal or Modula-2 program. As the program shown in Figure 8.12 illustrates, the reuse of names can cause confusion as to which object is being referenced. It is very easy to assume in assigning the value 25 to Y in PROCEDURE L4 that the global variable Y is being referenced, when in fact it is the variable Y declared in PROCEDURE L1 that is being referenced.

```
MODULE ScopeL2;
VAR   X,Y,Z :  INTEGER;
PROCEDURE L1;
     VAR   Y :  INTEGER;
     PROCEDURE L2;
          VAR X :  INTEGER;
          PROCEDURE L3;
               VAR Z :  INTEGER;
               PROCEDURE L4;
                    BEGIN
                         Y := 25;  (* L1.Y NOT L0.Y *)
                    END L4;
               BEGIN
                    (* L1.Y, L2.X, L3.Z visible *)
               END L3 ;
          BEGIN
               (* L1.Y, L2.X, L0.Z visible *)
          END L2 ;
     BEGIN
          (* L0.X, L1.Y, L0.Z visible *)
     END L1 ;
BEGIN
     (* ... *)
END ScopeL2.
```

Fig. 8.12   Loss of visibility in nested procedures.

8.10 MODULARITY

Both the FORTRAN-type languages and the Pascal-like block-structured languages provide some control over the visibility of names but the mechanisms provided are limited. Two particular problems arise in block-structured languages. The first is that there is no mechanism, other than reusing the name, to prevent an inner block accessing an object named in an outer block. A second problem arises because of the way in which storage is allocated – a locally declared object has storage dynamically allocated when the block is entered and the storage is de-allocated when the block is left – hence it is not possible to preserve the value of an object from one activation of the block to the next. As an example consider a procedure which is used to store characters in a buffer and another which is used to remove the characters:

```
PROCEDURE Put(character);
BEGIN
     buffer[i] := character;
     i := i+1;
END;
PROCEDURE Get(VAR character:CHAR);
BEGIN
     i := i-1;
     character := buffer[i];
END;
```

The objects named *buffer* and *i* must maintain their values from activation to activation of the procedures; hence in Pascal they must be declared as global since they are used by both procedures.

In order to support modularity and information-hiding, both of which contribute to security, a language must provide a mechanism for controlling explicitly the visibility of all names, including the ability to make local names available outside the unit in which the name is declared; it must also provide a means by which a local variable can maintain its value from one activation of the unit to the next. These problems have been addressed by a number of language designers and program units which provide the facilities that have been devised. They use several different names: a class in SIMULA, a module in Modula-2, a segment in ALGOL and a package in Ada.

The facilities provided in Modula-2 are divided into modules and local modules. Modules support both information-hiding and separate compilation: local modules support information-hiding but are not separately compilable. We will consider first the structure of a local module to illustrate the information-hiding and scope-control features of the module construct.

In Modula-2 the main program unit is a module and hence a local module is a module which is nested within the main module – note that further modules can be nested within the nested local module. The nesting of modules is illustrated in Figure 8.13.

In order to allow objects which are declared within the body of a module to be visible outside the module they must be listed in an EXPORT list. Similarly objects which are declared outside the module must be specifically imported into the module by naming them in an IMPORT list. It should be noted that all objects can be imported and exported, i.e., variables, constants, types and procedures. Objects which are declared in a module are created at the initialization of the program and remain in existence throughout the existence of the program, i.e., they are not like objects created inside a procedure which cease to exist when the procedure body is left.

The more general concept of a module allows a program to be split into many separate units, each of which can be compiled separately. The facilities for separation also allow for the construction of program libraries in which the library

```
MODULE ImportExport;
(*
Title    : Example of Import and Export of objects
File     : sb1 Importex.mod
LastEdit:
Author   : S. Bennett
*)
VAR a,b,c : INTEGER;
(*   ...    *)
  MODULE L1;
    IMPORT a;
    EXPORT d,g;
    VAR d,e,f : INTEGER;
      MODULE L2;
        IMPORT e,f;
        EXPORT g,h;
        VAR g,h,i : INTEGER;
        (*    ...
           e,f,g,h,i visible here
           ...   *)
      END L2;
    (*   ...
        a,d,e,f,g,h are visible here
        ...      *)
  END L1;
(*  ...
    a,b,c,d,g are visible here
    ...*)
END ImportExport.
```

Fig. 8.13 Scope and visibility control in Modula-2.

segments are held in compiled form rather than as source code. In order to do this the module is split into two parts:

1. DEFINITION MODULE – contains information about objects which are exported from the module.
2. IMPLEMENTATION MODULE – this is the body of the module which contains the code which carries out the functions of the module.

The *definition* part is made available to the client program in source form, but the *implementation* part is only provided in object form and its source code remains private to the module designer. The separation in this way provides an excellent method of hiding implementation details from a user and the actual implementation can be changed without informing the user providing that the definition part does not change. An example of a definition module is given in Figure 8.14 and further examples are given in Chapter 9.

8.11 INDEPENDENT AND SEPARATE COMPILATION

One of the reasons for the popularity and widespread use of FORTRAN for engineering and scientific work is the fact that sub-routines can be compiled

```
DEFINITION MODULE Buffer;
(*
Title    : Example of definition module
File     : sb1 Buffer.mod
LastEdit:
Author   : S. Bennett
*)
EXPORT QUALIFIED
     put, get, nonEmpty, nonFull;
VAR
     nonEmpty, nonFull : BOOLEAN; (* used to test the status of buffer *)
PROCEDURE put(x : CARDINAL);
     (* used to add items to the buffer *)

PROCEDURE get(VAR x : CARDINAL);
     (* used to remove items from the buffer *)
END Buffer.
```

Fig. 8.14 Example of a DEFINITION MODULE in Modula-2.

independently from the main program, and from each other. The ability to carry out compilation independently arises from the single-level scope rules of FORTRAN: the compiler makes the assumption that any object which is referenced in a sub-routine, but not declared within that sub-routine, will be declared externally and hence it simply inserts the necessary external linkage to enable the linker to attach the appropriate code. It must be stressed that the compilation is independent, i.e., when a main program is compiled the compiler has no information available to it which will enable it to check that the reference to a library sub-routine is correct. For example, a sub-routine may expect three real variables as parameters; if the user supplies four integer variables in the call statement the error will not be detected by the compiler. Independent compilation of most block-structured languages is even more difficult and prone to errors in that arbitrary restrictions on the use of variables have to be imposed.

Both Modula-2 and Ada have introduced the idea of separate compilation units. By separate, rather than independent, compilation is meant that the compiler is provided with some information about the separately compiled units which are to be incorporated into a program. In the case of Modula-2 the definition part of a separately compiled module must be made available to the user, and hence the compiler; this enables the compiler to carry out the normal type checking and procedure parameter matching checks. Thus in Modula-2 type mismatches and procedure parameter errors are detectable by the compiler. It also makes available the scope control features of Modula-2. The provision of independent compilation aids the development process through the ability to create object code libraries; the provision is enhanced if separate compilation is available in that it also improves the inherent security of the language.

8.12 EXCEPTION HANDLING

One of the most difficult areas of program design and implementation is the handling of errors, unexpected events (in the sense of not being anticipated and hence catered for at the design stage) and exceptions which make the processing of data by the subsequent segments superfluous, or possibly dangerous. One of the functions of the run-time support system is to trap errors which have not been anticipated. The typical sort of error which will be trapped by the run-time system is an attempt to divide by zero. The normal response is to halt the program and display an error message on the user's terminal.

In a development environment it may be acceptable for a program to halt following an error: in a real-time system every attempt must be made to keep the system running. Versions of languages such as BASIC, FORTRAN and Pascal intended for use for real-time systems have been produced in which the run-time error trapping mechanism can be set to pass control to a routine provided by the system designer when a run-time error is detected. The error can be checked and action taken to keep the system running: if it cannot be kept running then at least there may be an opportunity to close it down safely and warn the operator.

* * *

Example 8.3

Consider the boiler control system described in Chapter 2. One of the control loops uses the ratio of fuel flow and air flow to calculate the set points for the controller. Assuming that the values of fuel flow and air flow are read from the measuring instruments and stored in integer variables fuelFlow and airFlow respectively, the instruments provide values which are in the range 100 ... 4096 corresponding to the flow ranges 0 ... 100%. The control setting is to be held in a real variable, ratioSetPoint. The programmer might write the statement:

```
ratioSetPoint := fuelFlow/airFlow
```

A program using this statement may function correctly for a long period, but suppose at some time a fault on either the air-flow measuring instrument or the device used to convert the reading to an integer value results in airFlow being set equal to zero. The program, and hence the system, would halt.

One solution would be to validate the data prior to executing the statement with, e.g., the following:

```
IF airFlow > 0 THEN
    ratioSetPoint := fuelFlow/airFlow
ELSE
    airFlowAlarm := TRUE;
```

In this simple example it is relatively easy to put in the necessary checks. But checking for all possible data errors can become very complex, can obscure the general flow of the code, and can slow down execution. Use of exception handlers or

Exception Handling

error trap routines can simplify the flow of the code and group the error analysis into one place.

* * *

Example 8.4

Referring to the previous example we will assume that the values of air flow and fuel flow read from the instruments are the raw values (rawFuelFlow, rawAirFlow) and that these have to be converted to the actual values by taking the square root of the raw value; the code is thus:

```
fuelFlow := sqrt(rawFuelFlow);
airFlow : = sqrt(rawAirFlow);
ratioSetPoint := fuelFlow/airFlow
```

We now have two additional sources of error; if, through an instrument error, either rawFuelFlow or rawAirFlow is recorded as negative values then program execution will stop with an error message informing us that we have tried to obtain the square root of a negative number. We can, as in Example 8.3, use IF...THEN...ELSE statements to check each value, or we could use the ON ERROR type of construction supported by some language implementations. For example Texas Instruments Pascal provides a routine ONEXCEPTION(exceptionHandler:INTEGER) which informs the run-time system of the address of an exception handler to be invoked in the event of an error.

```
            BEGIN
                ONEXCEPTION(flowControlErrors);
                fuelFlow := sqrt(rawFuelFlow);
                airFlow : = sqrt(rawAirFlow);
                ratioSetPoint := fuelFlow/airFlow;
            END;
    PROCEDURE flowControlErrors;
    BEGIN
        IF rawFuelFlow < 0 THEN
            Alarm('rawFuelFlow negative');
        IF  rawAirFlow < 0 THEN
            Alarm('rawAirFlow negative');
        IF rawAirFlow = 0 THEN
            Alarm('rawAirFlow zero');
    END;
```

The obvious advantage of this approach is that readability of the code is improved – the designer can consider the normal function and the error actions as two separate problems – and execution speed is improved. The disadvantage is that it encourages reliance on the run-time system rather than careful consideration of the possible errors.

The use of error-trapping or exception statements requires careful consideration of the action which follows the execution of the error-handling routine. In Example 8.4 it is assumed that execution resumes at the next statement after the statement which generated the error. (How would the procedure 'flowControlErrors' have to been changed if execution was resumed at the statement causing the error?) Some languages, in particular BASIC, allow explicit control over the point at which execution is resumed either by allowing the use of a GOTO statement or by a RESUME statement which takes arguments specifying at which statement execution is to resume.

The Texas Pascal and Ada allow the user to raise exceptions, thus in Ada the program fragment shown in Example 8.4 could be written in the following way:

```
BEGIN
    IF rawFuelFlow < 0 THEN
        RAISE fuelError;
    fuelFlow := sqrt(rawFuelFlow);
    IF rawAirFlow <= 0 THEN
        RAISE airError;
    airflow : = sqrt(rawAirFlow);
    ratioSetPoint := fuelFlow/airFlow;
EXCEPTION
    WHEN fuelError =>
        ALARM(fuel);
    WHEN airError =>
        ALARM(air);
END;
```

The advantage of this approach is that it still separates the code dealing with the exception from the main body of the program but, because the designer has explicitly to raise exceptions, he has to consider possible errors and make provision for the errors.

The major problem in error and exception handling occurs when procedures are nested and it is required to transmit knowledge of the error from one procedure to another.

* * *

Example 8.5

Consider the following system

```
PROCEDURE A;
BEGIN
  ...
  B; ------- PROCEDURE B;
              BEGIN
                ...
                C; ---------PROCEDURE C;
                            BEGIN
                              ONEXCEPTION(flowControlErrors);
                              fuelFlow := sqrt(rawFuelFlow);
                              airFlow : = sqrt(rawAirFlow);
                              ratioSetPoint := fuelFlow/airFlow;
                  ----------  END;
                ...
  ---------   END;
  ...
  (* value of airFlow
     is to be used here *)
END;
```

Low-level and Multi-tasking Facilities 301

If the value of rawAirFlow is found to be in error in Procedure C then the code in Procedure B and in Procedure A is not relevant since it is assumed to need a correct value for rawAirFlow. The only mechanism available in languages such as Pascal is to use a BOOLEAN variable to indicate if the value is valid; the variable will have to be either a global variable or passed as a parameter from procedure to procedure.

The solution which has been adopted in Ada is that when an exception is raised it is passed from block to block until the appropriate exception handler is found. Thus in the above example, if the exception was raised in Procedure C and there was no exception handler in C, then C would be terminated and control would pass to B; if there were no handler in B it would also be terminated and control passed to A.

* * *

8.13 LOW-LEVEL AND MULTI-TASKING FACILITIES

In programming real-time systems there is frequently a requirement to be able to directly manipulate data in specific registers in the computer system: physical memory registers; CPU registers; and registers in an input/output device. The majority of the high-level languages provide no facilities to do this other than the use of assembly coded routines. In weakly-typed languages there is little difficulty in extending the language to allow an integer to be treated as a fixed length bit string and to provide a mechanism by which it can be mapped on to a memory register or other machine register.

The simplest mechanism is the use of procedures (or sub-routines) and functions and many microprocessor versions of FORTRAN and BASIC provide the following:

PEEK(address) – returns as INTEGER variable contents of the location address.
POKE(address, value) – puts the INTEGER value in the location address.

It should be noted that on 8-bit computers the integer values must be in the range 0 .. 255 and on 16-bit machines they can be in the range 0 .. 65535. For computer systems in which the input/output devices are not memory mapped, e.g. Z80 systems, additional functions are usually provided such as INP(address) and OUT(address, value).

A slightly different approach has been adopted in BBC BASIC which uses an 'indirection' operator. The indirection operator indicates that the variable which follows it is to be treated as a pointer which contains the address of the operand rather than the operand itself (the term 'indirection' is derived from the indirect addressing mode in assembly languages). Thus in BBC BASIC the following code

```
100 DACAddress = &FE60
120 ?DACAddress = &34
```

results in the hexadecimal number 34 being loaded into location FE60H; the indirection operator is '?'. The use of the indirection operator is illustrated in the program given in Figure 8.3.

In some of the so-called Process FORTRAN languages and in CORAL and

RTL/2 additional features which allow manipulation of the bits in an integer variable are provided, e.g., SET BIT J(I), IF BIT J(I), n1,n2 where I refers to the bit in variable J. Also available are operations such as AND, OR, SLA, SRA, etc., which mimic the operations available at assembly level.

The weakness of implementing low-level facilities in this way is that all type-checking is lost and it is very easy to make mistakes. A much more secure method would be to allow the programmer to declare the address of the register or memory location but to also be able to associate a type with the declaration:

```
VAR charout AT 0FE60H :CHAR;
```

which declares a variable of type CHAR located at memory location ∅FE6∅H. Characters can then be written to this location by simple assignment

```
charout := 'a';
```

Note that the compiler would detect and flag as an error an attempt to make the assignment

```
charout := 45;
```

since the variable is typed.

Both Modula-2 and Ada permit declarations of the above type and an example of a Modula program which uses the low-level facilities is given in Example 8.6. This program illustrates a further requirement for low-level support from the high-level language: the ability to handle interrupts. In the example shown SuspendUntilInterrupt is a high-level procedure provided by the module Processes which is part of the NUPD RTS kernel (see Chapter 9).

*   *   *

Example 8.6

```
MODULE TermOut[4 (* interrupt priority of device *)];
  FROM PROCESSES IMPORT
    SuspendUntilInterrupt;
  EXPORT
    PutC;
  CONST
    readyBit = 7;
    interruptEnableBit = 6;
    interruptVector = 64B;
  VAR
    ttyReg[177564B] : BITSET;
    ttyBuf[177566B] :CHAR;
  PROCEDURE
    PutC(c: CHAR);
  BEGIN
    IF NOT(readyBit IN ttyReg) THEN
      INCL(ttyReg, InterruptEnableBit);
      (* high processor priority will fend off
      the interrupt until ...*)
      SuspendUntilInterrupt(interruptVector);
      EXCL(ttyReg, interruptEnableBit);
    END;
    ttyBuf := c;
  END PutC;
END TermOut;
```

This example also shows how Modula-2 handles bit-level manipulation. It is possible to declare a register as of type BITSET and perform set operations on the register. The operators INCL and EXCL are respectively the operations of including a bit in the set, i.e. setting a bit in the register, and excluding a bit from the set, i.e. resetting a bit in the register.

* * *

Modula-2 provides a low-level support mechanism for interrupts through the procedure

```
PROCEDURE IOTRANSFER(VAR source, destination:PROCESS;
                    va:CARDINAL)
```

which is inserted in the interrupt routine. The statement causes control to be passed to a named co-routine – the destination process – and on receipt of an interrupt control is returned to the source process at the statement following the IOTRANSFER statement.

The support for multi-tasking in real-time languages takes one of two forms: linkage to operating system routines as described in Chapter 6 and provision of support mechanisms within the language as described in Chapter 7.

EXERCISES

8.1 Examine the list of run-time errors generated by any language system which you use, or for which you have a guide. List the error conditions for which you think the system could either recover or be closed down in a safe manner. Which conditions would require a resort to the run-time support error handling?

8.2 Write an input routine which reads a date from a terminal. The data has to be expressed in the form dd:mm:yyyy, where dd = day of the month, mm = number of the month (1 .. 12) and yyyy = the year (range 1980 .. 2000). The routine must trap all possible errors.

8.3 Rewrite the routine in the previous exercise on the assumption that the programming language has an exception mechanism which passes an error code to a named routine.

8.4 Rewrite the program given in Figure 8.1 in a structured language such as Pascal or Modula-2. Which of the variables given in the program would you change into constants and why?

REFERENCES AND BIBLIOGRAPHY

ASHCROFT, E. and MANNA, Z. (1971), 'The translation of "go to" programs to "while" programs', *Proceedings of the 1971 IFIP Congress* **1** (North Holland Publishing 1972), pp. 250–5, also in Yourdon (1979), pp. 51–64

BARNES, J.G.P. (1976), *RTL/2 Design and Philosophy*, Heyden

DAHL, O.J., DIJKSTRA, E.W. and HOARE, C.A.R. (1972), *Structured Programming*, Prentice Hall

DIJKSTRA, E. (1968), 'Go to statement considered harmful', Letter to Editor, *Communications of ACM*, **11(3)**, pp. 147–8, also in Yourdon (1979) pp. 29–33

FISHER, D.L. (1983), 'Global variables versus local variables', *Software: Practice and Experience*, **13**, pp. 467–9

GAIT, J. (1983), 'A class of high-level languages for process control', *Computers in Industry*, **4(1)**, pp. 63–7

HENRY, R. et al. (1985), *The ModOS User's Manual*, Human-Computer Interaction Group, Psychology Department, Nottingham University

HOPKINS, M.E. (1972), 'A case for the GOTO', *Proceedings of the 25th National ACM Conference*, 1972, pp. 791–7, also in Yourdon (1979) pp. 101–12

HOYLE, B.S. (1984), 'Engineering microprocessor software', *Electronics and Power*, **30(5)**, pp. 395–9

KNUTH, D.E. and FLOYD, R.W. (1971), 'Notes on avoiding 'GO TO' statements', *Information Processing Letters*, **1**. pp. 23–31, also in Yourdon (1982), pp. 153–62

MORALLEE, D. (1984), 'Programming languages: Where next?', *Electronics and Power*, **30(5)**, pp. 400–5

PRESSMAN, R.S. (1982), *Software Engineering: a Practitioner's Approach*, McGraw-Hill

VINCENT, G. and GILL, J. (1981), *Software Development*, Texas Instruments

WIRTH, N. (1982), *Programming in Modula-2*, Springer

WULF, W.A. (1972), 'A case against the GO TO', *Proceedings of the 25th National ACM Conference*, 1972, pp. 791–7, also in Yourdon (1979) pp. 85–100

YOUNG, S.J. (1982), *Real-time Languages: Design and Development*, Ellis Horwood

YOUNG, S.J. (1983), *An Introduction to Ada*, Ellis Horwood

YOURDON, E.N. (1979), *Classics in Software Engineering*, New York, Yourdon Press

YOURDON, E.N. (1982), *Writings of the Revolution*, New York, Yourdon Press

YOURDON, E.N. (1986), *Managing the Structured Techniques*, New York, Yourdon Press (1st edition 1976)

9

Programming Languages

9.1 ASSEMBLY LANGUAGES

There will always be areas in which the use of assembly languages is essential. Assembly languages are required when very tight time and space constraints are placed on the software. Typical application areas are in the aerospace and defense industries, in which the time constraints usually predominate, and in the mass market consumer industries, where it is space that is important.

Assembly languages obviously provide excellent low-level facilities with full access to the computer hardware. Their weaknesses are the lack of any data typing or data structures, poor readability, limited control structures and no explicit multi-tasking support. The readability of assembler code is poor because the language relates directly to the CPU and memory registers, and not to application objects. For example, a typical assembly instruction is LD A, (HL) which translated means: load into the register A the contents of the location whose address is stored in the register pair HL. The programmer has to remember, or record as a comment, the object whose address has been placed in HL.

The maintainability of assembly-coded software is heavily dependent on good documentation and on such documentation being kept up-to-date. It should also be remembered that because assembly code is directly related to a specific processor family it is not portable.

The ease of use of assembler is very dependent on the support facilities: good support facilities will provide for program modularization with separate assembly of segments and scope control, construction of libraries and interactive debugging facilities (including the ability to study in detail the timing of the execution of the code). Powerful editing facilities, including a multiple window facility to make reuse of code segments easy, and version control, are essential for efficient use of assembly code.

Most of the major microprocessor chip manufacturers sell development systems which provide extensive support facilities for assembly-level coding, while some general purpose development systems can be 'personalized' to a range of different microprocessor chips. In applications in which extensive assembly level development is required the investment in a development system can very quickly be recovered through the reduction in development time. Development work can be enhanced if the development system also supports some high-level languages and makes the mixing of high-level and assembly level segments easy.

An example of one type of development system is given in Figure 9.1. Three

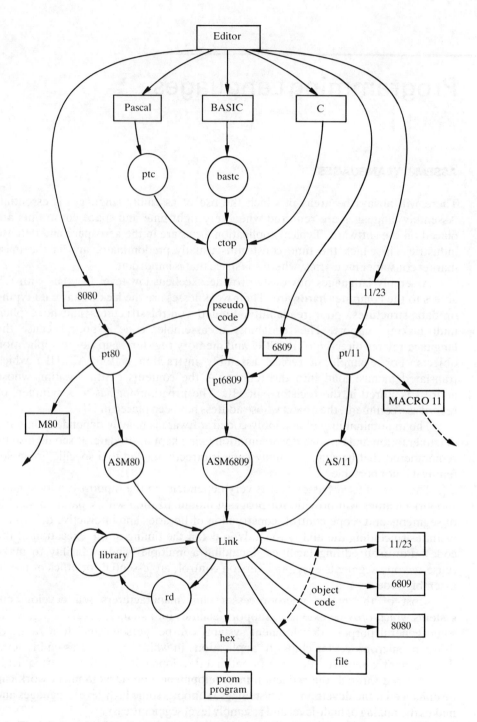

Fig. 9.1 Structure of a microprocessor development system.

high-level languages are supported: Pascal, BASIC and C. Program segments written in Pascal or BASIC are translated to C. The C code is compiled into an intermediate code called P code (pseudo code). The P code can be translated into a number of different assembly codes, three (8080, 6809 and PDP11) are shown in the figure. At this stage code segments written in the appropriate assembly language can, if required, be merged with the translated high-level language. The assembly code can either be assembled on the development system, linked and loaded into the appropriate processor module (or into the PROM programmer) or translated into a form in which it can be used by other standard assemblers, e.g., Microsoft M80, or DEC MACRO 11.

The other area in which assembly coding is sometimes required is for small parts of large systems. It is frequently the case that one part of a larger system may have an operation which is time-critical; it then becomes important to consider the ease with which an assembly routine can be linked to code written in the main programming language. Some languages allow assembly code to be inserted in line with the main language, others insist that any assembly language routines have to be supplied as separate external procedures or sub-routines which have to be merged with the main code when it is linked. The major problems which arise with mixing an assembly language and a compiled language are concerned with the sharing of data: with the compilers which use static storage allocation (e.g. FORTRAN) the problem is largely concerned with which part of the program allocates the storage – the assembler or the compiler – with dynamic allocation of storage the problem is more difficult; the usual solution is for the compiled code to allocate the storage and pass a pointer to the address via the machine registers. If the data to be shared can be contained in the machine registers, linkage is simple if the compiler allows explicit assignment of variable to specific registers.

The decision to use assembly code for a software project ought to be a positive decision. There should be definite, identifiable reasons why assembly code needs to be used and these reasons should be documented.

9.2 EVOLUTION OF HIGH-LEVEL LANGUAGES

The problems and cost of developing large systems using assembly coding has led to the use of a wide range of languages for real-time work and these can be categorized as follows.

1. *Low-level languages*

    (a) assemblers;
    (b) macro-assemblers;

2. *High-level languages*

    (a) Compiled

(i) FORTRAN;
(ii) COBOL;
(iii) PL/1;
(iv) Coral 66;
(v) RTL/2;
(vi) C;
(vii) Pascal;
(viii) Modula-2;
(ix) Ada;
(b) Interpreted
(i) BASIC;
(ii) FORTH;

3. Application languages

(a) APT (for numerically controlled machine tools);
(b) CUTLASS (language for process control);
(c) Conic (a language for distributed computing).

The languages can be classified in different ways, e.g. Pressman [1982] divides them into:

1. FOUNDATION
   (a) FORTRAN;
   (b) COBOL;
   (c) ALGOL;
   (d) BASIC;
2. STRUCTURED
   (a) ALGOL;
   (b) PL/1;
   (c) PASCAL;
   (d) C;
   (e) Modula-2;
   (f) Ada;
3. SPECIALIZED
   (a) FORTH;
   (b) APT;
   (c) CUTLASS.

The foundation languages have a very wide usage, extensive software libraries, are well-known, and have gained wide acceptance. Despite the widespread (and justified) criticism of it, FORTRAN is the major programming language for engineering and scientific use. Similarly COBOL is the major language for commercial data processing.

BASIC was designed as a simple language to teach programming to beginners in a time-sharing environment. In its original form BASIC was a cut-down version of

Evolution of High-level Languages

FORTRAN, but through its command features, i.e., the ability of the user to interact with programs and program fragments, it introduced the idea of a programming support environment. It has proved to be enormously popular and has gained a large user base with the development of personal computers. There are, however, hundreds of different versions with no standardization which makes portability difficult. As an interpretative language it makes the development of small systems quick and easy.

ALGOL was the forerunner of the structured languages and introduced the ideas of block-structuring, dynamic storage allocation and recursion which have become standard features of the structured programming languages. It has been widely used in Europe, but not so widely used in the USA. Some indication of the influence of ALGOL can be gained from Figure 9.2 which shows the general development of some of the main computer languages. It can be seen that the structured languages are descendants of ALGOL.

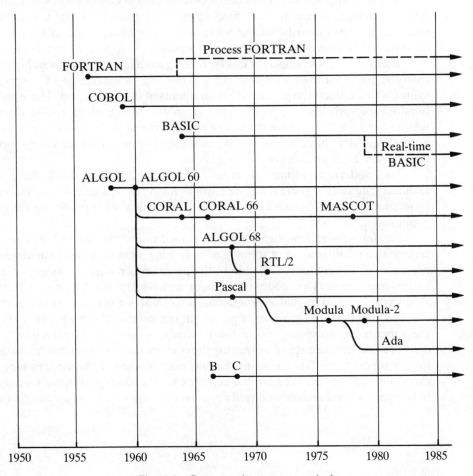

Fig. 9.2 Program language evolution.

Pascal was originally designed, like BASIC, as a language for the teaching of programming. It has also gained a wide user base in engineering and scientific work. The wide use has exposed some of the limitations of the language – in particular its lack of facilities to support compiled libraries – and led its designer (Wirth) to produce Modula and Modula-2. PL/1 was the first structured language to attempt to provide general support for all applications: commercial, engineering, scientific, and systems programming (support for I/O and multi-tasking). The language C was developed as a systems programming language, but is gaining acceptance outside that field.

Two languages were developed in Britain in the late 1960s and early 1970s: CORAL 66, which was developed for the defense based industries and which became a mandatory requirement for many defense contracts; and RTL/2 which was developed by Barnes at ICI and after extensive use within ICI was marketed commercially. Both languages drew heavily on ALGOL.

Ada is a language which was developed for the US Department of Defense as a new standard language for the programming of embedded real-time computer systems. It draws heavily on the ideas which have been used in Pascal and its derivatives and provides a powerful set of features to support real-time operations.

The applications languages can be considered as being at an even higher level than the so-called high-level languages. They are characterized by a specialized syntax which reflects the particular requirements of the application. There are many hundreds of such languages but each has only a small user base. The difficulties which arise with their use are the lack of portability and problems of maintainability. Some, e.g. APT, have gained a wide user base and international acceptance with defined standards and support.

The modern real-time languages such as Modula-2 and Ada have all the facilities required for general purpose programming as well as real-time and systems programming and it is expected that they will gain a wide user base for general applications.

The foundation languages were developed to enable non-technical programmers with a small amount of training to code what were by today's standards small programs. The only tools available to the programmer were the language manual and coding sheets. The modern languages such as Modula-2, C and in particular Ada were designed to enable programmers to construct very large programs and it was assumed that a large range of programming tools would be provided as part of the support environment. These tools include context-sensitive editors, version control software, debugging and testing support, module creation, file management, library support, etc. The criticism that has been leveled at the modern languages, especially Ada, is that they are too complex for the average programmer and that only highly trained and experienced programmers will be able to use them safely.

9.3 BASIC

BASIC (beginners all-purpose symbolic instruction code) was developed as a language which would provide a reasonably powerful range of facilities but which would be easy for the novice to learn and use. In particular the use of the interpretative mode of implementation was intended to make the writing, running and debugging of programs as quick and easy as possible. There is little doubt that BASIC has achieved the designer's aims and its widespread availability on microprocessor systems has also contributed to its use.

There are now so many versions of BASIC with numerous extensions to the original language that general comment is difficult. The original simple form of BASIC lacks most of the features considered necessary for a language for implementing real-time systems – it has poor data structuring, poor control structures, poor readability, no scope control (all objects are global), no low-level facilities, no multi-tasking facilities and, because it is interpreted, slow execution time. Some of the features which are lacking in the original BASIC have been incorporated in, e.g., BBC BASIC which provides:

- long variable names, using both upper and lower case letters;
- REPEAT .. UNTIL;
- WHILE .. DO;
- IF .. THEN .. ELSE;
- procedures with local variables;
- chaining of program segments;
- timing; and
- in-line assembly coding.

An example of code written using BBC BASIC was shown in Figure 8.4.

Versions of BASIC which have added simple low-level facilities and some means of timing are very suitable for certain types of real-time systems. In particular, BASIC is a useful language for experimental and development work where the speed and ease with which programs can be written and debugged is a great advantage; it is also useful for one-off small systems (say less than 1000 lines of code) which do not require high security of operation. It is not suitable for medium to large systems.

The simplest extensions to BASIC to provide for embedded system use are the provisions of low-level access mechanisms, e.g. PEEK, POKE and the indirection operators which were described in section 8.14.

Other extensions include built-in functions for handling input and output to analog-to-digital and digital-to-analog converters (see Mellichamp), event handling (interrupts) and multi-tasking.

The simplest form of event handling allows the use of statements of the form

```
ON EVENT GOSUB <n>
```

where ⟨n⟩ is a specified line number. This may be extended in some BASICs to allow several different events:

```
ON INTERRUPT0 GOSUB <n1>
ON INTERRUPT1 GOSUB <n2>
ON TIMEOUT GOSUB <n3>
```

The GOSUB is used so that a RETURN statement can be used to indicate the end of the section of code for the particular event.

An example of a more extended system is Multi-BASIC, which provides a real-time BASIC with up to eight tasks for the BBC microprocessor. In this system each task body is indicated by statements

```
TASK <name>
...
EXIT
```

which are used to bracket the statements which form the task. The task remains dormant (in terms of the task states described in Chapter 6, existent but not active) until an enable statement is executed.

```
ENABLE <name> EVERY 10cs
```
or
```
ENABLE <name> WHEN event
```

The tasks can be disabled by the statement DISABLE (name).

An example of a program with two tasks is shown in Figure 9.3. The two tasks are the simulation of a simple plant and control using a simple controller as is shown in Figure 9.4.

As was discussed in section 8.3.6, there is a temptation when using BASIC to reduce readability in order to increase the efficiency both in terms of the amount of memory used and in speed of execution. Most suppliers provide guidance on how to do this for their interpreter and some recommend having two copies of the program, one in which there are comments and the other, the code which is run, without comments. Using this method requires strong discipline to ensure that the actual code in each copy is identical and that when changes are made they are made in both copies.

The major advantage of BASIC is the ease with which the language can be learnt. If well-trained programmers with experience of other languages are not available then BASIC may be a very sound choice and result in better, more secure software, than could be achieved by using an inherently more secure language.

9.4 FORTRAN AND PASCAL

The standard versions of both FORTRAN and Pascal lack any real-time facilities. For both languages, real-time versions which have been produced by a range of computer manufacturers and software houses exist. The real-time features have been added as extensions to the language and the lack of standardization reduces the portability of the software. There is, in fact, a standard for real-time FORTRAN

```
10 REM   MULTI-TASKING USING MULTI BASIC
20 REM  ************************************************************
30 REM
40 REM S BENNETT JULY 1986
50 REM
60 REM  ************************************************************
70 REM
80 A$="Z3SYS"   :REM FILE NAME
90 REM
100 DIM RES(100)
110 INPUT "POINTS";POINTS%
120 REM
130 REM TASK DEFINITION FOR SIMULATION OF PLANT *****************
140 REM
150 TASK simul
160 UN2=UN1
170 UN1=UN
180 CN2=CN1
190 CN1=CN
200 CN=AA*UN+B*UN1+C*CN1-D*CN2
210 time=time+T
220 EXIT
230 REM
240 REM END OF TASK simul ******************************************
250 REM
260 REM TASK DEFINITION FOR CONTROL TASK ***************************
270 REM
280 TASK control
290 EN=RN-CN
300 UN=(K/A*A)*EN
310 RES(J)=CN
320 J=J+1
330 EXIT
340 REM
350 REM END control ************************************************
360 REM
370 REM MAIN ******************************************************
380 REM
390 K=2
400 A=1
410 T=0.02
420 REM
430 REM CALCULATE CONSTANTS
440 REM
450 D=EXP(-A*T)
460 AA=T*A-1+D
470 B=1-D-T*A*D
480 C=1+D
490 REM
500 REM INITIAL CONDITIONS
510 REM
520 UN=0
530 CN=0
540 UN1=0
550 UN2=0
560 CN1=0
570 CN2=0
580 time=0
590 RN=1
600 J=0
610 ENABLE simul EVERY 2cs
620 ENABLE control EVERY 20cs
630 REPEAT UNTIL J>=(POINTS%-1)
```

Fig. 9.3 Example of multi-tasking BASIC.

```
640 DISABLE simul
650 DISABLE control
660 REM
670 REM CALCULATE MAXIMUM VALUE OF RESPONSE
680 REM
690 FOR J=0 TO (POINTS%-1) STEP 2
700   PRINT RES(J);TAB(20);RES(J+1)
710 NEXT J
720 MAX=RES(0)
730 FOR J=0 TO (POINTS%-1)
740 IF RES(J)>MAX THEN MAX=RES(J)
750 NEXT J
760 PRINT "MAXIMUM VALUE=";MAX
770 INPUT "WAIT",N
780 REM
790 REM DRAW GRAPH OF RESPONSE
800 REM
810 CLG
820 MOVE 280,0
830 DRAW 280,1024
840 MOVE 0,24
850 DRAW 1280,24
860 MOVE 280,500
870 DRAW 1280,500
880 MOVE 280,24
890 X%=280
900 FOR J=0 TO (POINTS%-1)
910   Y%=RES(J)*500+24
920   X%=X%+(1000 DIV POINTS%)
930   DRAW X%,Y%
940 NEXT J
950 STOP
```

Fig. 9.3   (cont.)

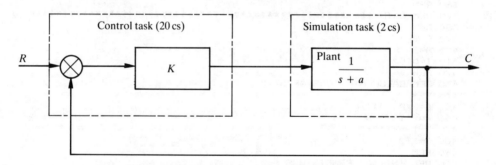

Fig. 9.4   System – program for which is shown in Figure 9.3.

which has been produced by the ISA (Instrument Society of America). The ISA standards cover:

1. Executive interface routines.
2. Process input/output routines.
3. Bit string functions.
4. Random file handlers.
5. Task management routines.

The extensions involve the specification of external procedure calls rather than a change to the syntax of the language.

The advantage of using FORTRAN is that it is a very widely-known language in the engineering community, extensive sub-routine libraries exist and it can be compiled independently. The disadvantages are that it has weak typing, limited control constructs and the scope control is clumsy to use, leading to errors in large programs. Many of the problems and criticisms which relate to FORTRAN II and FORTRAN IV have been corrected in FORTRAN 77. The weaknesses which remain are poor data structure support and weak typing.

Standard Pascal is very poorly suited for real-time control and large system work. A major problem is the lack of independent compilation; this slows down development work and also means that program libraries have to be distributed in source code form. Distribution in source code form, as well as the obvious commercial disadvantages, has the disadvantage that it contributes to poor security: it is difficult to change object form library routines, but it is very easy to change source code and such changes may result in errors appearing some time after the change has been made. Such error will be difficult to find since it is natural to assume that the library routines are correct.

A number of vendors supply Pascal compilers which support independent compilation, but this is usually subject to very tight constraints on object declaration. Because independent compilation is non-standard, portability is lost since the various vendors use different approaches.

Pascal has proved to be a highly popular language, particularly for the teaching of programming concepts. Its popularity has led to a number of 'enhanced' versions: these include, e.g., Concurrent Pascal – a language developed for experiments in concurrent programming – and a range of 'Pascals' developed by microprocessor vendors for real-time applications. The real-time versions have added multi-tasking facilities, signals and semaphores, and access to the computer hardware through the ability to manipulate variables containing memory addresses. The extensions take Pascal closer to Modula-2 and it is difficult to find any justification for using the enhanced Pascal instead of Modula-2.

9.5 CORAL 66

CORAL 66 was developed in the late 1960s at the Royal Radar Establishment, Malvern, England. It has for a number of years formed the standard language for defense-related systems. The language, which was derived from ALGOL 60, fails to address the main problems relating to a language for real-time applications, but this is not surprising since the problems were not well understood at the time of its development.

The distinctive features of CORAL 66 are its ability to handle bit manipulations, absolute addresses and fixed point arithmetic operations. The bit manipulations and absolute addressing provide access to the machine functions. The usefulness of such operations in CORAL 66 is, however, limited in that the language does not support parallelism and hence it is not feasible to handle interrupts from within the language.

The provision of fixed-point arithmetic support in CORAL 66 reflects the concentration of the language designers on providing an efficient language in both speed of execution and size of generated code. Again this reflects the time when the language was designed and its purpose. Support for fixed-point arithmetic operations has always been a difficult area for language design and Young [1982, p. 57] reports that the CORAL 66 facility is so mistrusted by programmers that it is rarely used. There is much less emphasis on fixed-point arithmetic operations now that computers are faster and hence floating point operations performed using software routines are reasonably quick. Also, it is no longer prohibitively expensive to use hardware floating-point arithmetic units.

For computational efficiency CORAL 66 uses static allocation of data storage rather than the dynamic allocation which might be expected of a derivative of ALGOL.

The language itself does not support multi-tasking: if multi-tasking is required it must be provided by the underlying operating system. The problems this has caused, through a lack of standardization between different compilers, has led to the development of the MASCOT system which defines precisely the characteristics for multi-tasking support.

Exception handling is limited to a form of computed goto statement: an array of statement labels (called a SWITCH) can be generated by the statement

```
'SWITCH' X LAB1,LAB2,LAB3,LAB4
```

The statement

```
'GOTO' X(I)
```

causes a jump to the Ith label in the switch list. Any number of different switch arrays can be declared. It is illustrative of the poor security of the language that there is no check of the bounds of I.

In many ways CORAL 66 lies between the powerful macro assemblers and the more modern real-time high-level languages. It is perhaps for this reason that it appealed to programmers experienced in using assembly languages who liked the ease with which assembly statements could be included in the code and the macro feature supported by CORAL 66.

9.6 RTL/2

RTL/2 was designed by J.G.P. Barnes of ICI. The first implementation was completed in 1971; and in 1974, after extensive use within ICI, it was made commercially available. As a language it represents an important and serious response to the technical requirements of real-time programming. It provides some limited support for multi-tasking; it supports access to the machine hardware; has a module structure; but does not provide exception handling facilities.

In RTL/2 a program is made up of a number of 'bricks'. These can be procedure, data or stack bricks:

1. Procedure brick. This consists of pure code and is re-entrant; it may have parameters and local variables, but these are restricted to scalars. Procedures may not be nested.
2. Data brick. This is a named global static collection of objects; it is similar to FORTRAN named COMMON.
3. Stack brick. This is an area used as a workspace by a task for the storage of local variables etc.

A task is formed from a main procedure and a stack brick. Usually the main procedures will be different for the different tasks. Other procedures can be called from the main procedures and these may be shared between several tasks. Each task is assumed to run concurrently. Data bricks may be private to a task, or may be shared between one or more tasks; if they are shared access must be coordinated by use of semaphores. The general structure of a program showing several tasks is illustrated in Figure 9.5.

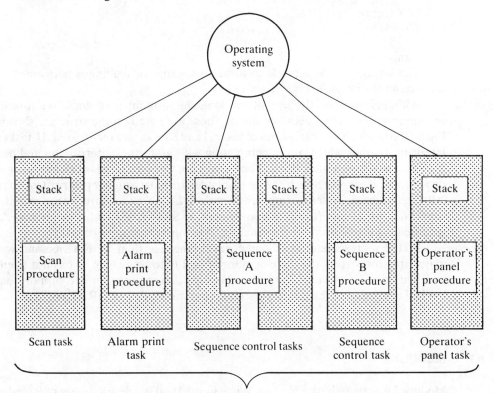

Fig. 9.5 General structure of an RTL/2 program.

Separate compilation is provided through the use of a MODULE which consists of a set of declaration statements and a number of bricks (procedure, data and stack). There is explicit control over the visibility of objects between modules. This is simple in the case of procedure and stack bricks: they are named in ENT and EXT statements which imply that they are to be exported (ENT) and imported (EXT). For data bricks the position is more complicated in that information of the type and number of object within the data brick has to be provided. Hence in the exporting module the declaration of the brick is preceded by the word ENT and, in the importing module, the whole declaration has to be repeated after the word EXT, with the objects being declared in the same order as in the exporting module. For example

```
ENT DATA TEST;
    INT N,M;
    REAL X;
ENDDATA;
```

would have to appear in the exporting module and the code

```
EXT DATA TEST;
    INT N,M;
    REAL X;
ENDDATA;
```

in the importing module.

The linkage of the module to form a program for loading is performed by a separate utility program known as a linker.

Although not strictly part of the language a number of small multi-tasking operating systems have been specified. These are based on the structure shown in Figure 9.6 in which two categories of task, H and S tasks, are recognized. H-tasks are interrupt routines which run to completion without further interruption and hence can share a common stack brick. The S-tasks are software tasks which are organized on a priority basis. An S-task can be suspended to allow a higher priority task to run. All the operating systems support a simple DELAY procedure; other primitives supported are events (these are equivalent to SIGNALS) and semaphores. Also specified is a mechanism for the creation, starting and stopping of tasks.

RTL/2 represents a much higher level language than CORAL 66 and was a major step forward in the development of real-time languages. It has not been so widely used as CORAL 66, but this is largely because of the limited support that it has had compared to the large investment that was made in CORAL 66.

9.7 MODULA-2

Modula-2 was specified by Niklaus Wirth in 1980 and is a development of Pascal and Modula. The first working compiler was released in 1981. A number of commercially

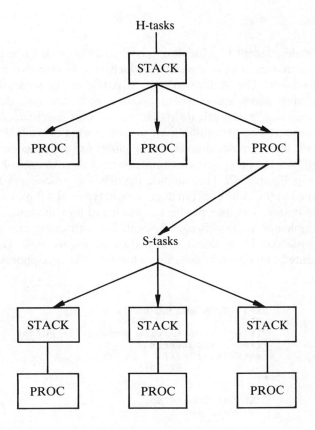

Fig. 9.6 H- and S-tasks in RTL/2.

produced compilers are now available and the British Standards Institution formally began work on producing a BSI language specification in January 1985.

As might be expected Modula-2 is very similar to Pascal but corrects a number of its shortcomings. The new features introduced are the LOOP ... EXIT construction, new standard TYPES (BITSET, CARDINAL, WORD and ADDRESS), and new functions (VAL, CAP, INC, DEC, ALLOCATE, SIZE, and ADR). As has been previously discussed, the LOOP ... EXIT construction simplifies the escape from a repetitive loop. The additional types and functions make it possible to use Modula-2 for low-level programming and avoid the need for the use of assembly code. The Pascal syntax is also tidied up by the consistent use of explicit termination symbols: this avoids the need for the frequent use of BEGIN ... END pairs to bracket code segments.

The two major changes from Pascal are, however, the MODULE concept and the provision of concurrent programming facilities.

9.7.1 Modules

The module concept of Modula-2 has already been described in Chapter 8. The basic compilation unit is a *module*, which is split into two parts: *definition* and *implementation*. The 'definition' part is public in the sense that any user of the module must have access to the source code for the 'definition' part. The 'implementation' part is private in that the user only requires access to object code in order to use it. The 'definition' part describes what is available in the module at a high level; for example, it gives the headings for the procedures available but no details of how the procedures are implemented. An example of a definition module is given in Figure 9.7. This module exports the procedures Put and Get which enable the user of a buffer to put data items of type CHAR into a buffer and get items from the buffer. The user has no knowledge of how the buffer is implemented. A buffer implemented as a first-in-first-out list with the operations of Put and Get being protected by a monitor is shown in Figure 9.8. The buffer could be implemented in a different way, using for example semaphores as shown in Figure 9.9.

```
DEFINITION MODULE Buffer;
(*
Title    : Definition module for buffer
File     : buffer.def
LastEdit : 17/3/7
Author   : S. Bennett
*)

   EXPORT QUALIFIED
      Put, Get ;

   PROCEDURE Put (ch : CHAR);

   PROCEDURE Get(VAR ch:CHAR) ;

END Buffer.
```

Fig. 9.7 Example of a Modula-2 DEFINITION module.

The use of IMPORT and EXPORT lists within modules allows the programmer explicit control over the visibility of objects. (The need to list objects to be exported in an EXPORT QUALIFIED list has now been removed from the definition of the language, but many existing compilers still require such a list.)

An advantage of splitting modules into two parts is that it is possible to create program libraries as when using FORTRAN, but because the compiler has available the information in the 'definition' module it can carry out type-checking and ensure the objects imported from the module are used consistently and according to their definition. As when using FORTRAN, the details of the implementation of the library routine are hidden from the user. A second advantage is that the internal code of the 'implementation' module can be changed without the need to change any of the client modules provided that the definition remains unchanged. (The modules

```
IMPLEMENTATION MODULE Buffer ;
(*
Title    : Implementation of a buffer using a monitor
File     : buffert1.mod
LastEdit : 17/3/87
Author   : S. Bennett
System   : LOGITECH MODULA-2/86 + RTS Kernel
*)
FROM Monitor IMPORT
     monitorPriority;
  FROM Semaphores IMPORT
    Claim, InitSemaphore, Release, Semaphore;
  (* following is required for display of contents *)
    FROM Ansi IMPORT
       WriteCh;
  (* end of display *)

CONST moduleName = 'bufferT1';


MODULE BufferM [monitorPriority];
IMPORT
   Claim, InitSemaphore, Release, Semaphore, WriteCh;

EXPORT Put, Get;
  CONST
       nMax = 10;

  VAR
     nFree, nTaken : Semaphore;
     in, out : [1..nMax];
     b: ARRAY [1..nMax] OF CHAR;


  (* following variables are required only for demonstration purposes *)
     row, col : CARDINAL;

     PROCEDURE Put (ch : CHAR);
     BEGIN
        Claim(nFree);
        b[in] := ch;
        in := in MOD nMax + 1;
         WriteCh(ch, row, col+in)  (* display purposes only *);
        Release(nTaken)
     END Put;

     PROCEDURE Get(VAR ch:CHAR) ;

     BEGIN
        Claim(nTaken);
        ch := b[out];
        out := out MOD nMax + 1;
         WriteCh(' ', row, col + out) (* display purposes *);
        Release(nFree)
     END Get;

     BEGIN
       row:=15; col:=20 (* initialise display part *);
       in:=1; out:=1;
       InitSemaphore(nFree, nMax);
       InitSemaphore(nTaken, 0);
     END BufferM;

  END Buffer.
```

Fig. 9.8   IMPLEMENTATION MODULE of buffer using a monitor.

```
IMPLEMENTATION MODULE Buffer ;
(*
Title   : Implementation of a buffer using a monitor
File    : buffert2.mod
LastEdit: 17/3/87
Author  : S. Bennett
System  : RTS Kernel
*)
   FROM Semaphores IMPORT
      Claim, InitSemaphore, Release, Semaphore;
   (* following is required for display of contents *)
      FROM Ansi IMPORT
         WriteCh;
   (* end of display *)

CONST moduleName = 'bufferT2';


   CONST
         nMax = 10;

   VAR
        nFree, nTaken, inPut, inGet : Semaphore;
        in, out : [1..nMax];
        b: ARRAY [1..nMax] OF CHAR;


   (* following variables are required only for demonstration purposes *)
        row, col : CARDINAL;

        PROCEDURE Put (ch : CHAR);
        BEGIN
           Claim(inPut);
           Claim(nFree);
           b[in] := ch;
           in := in MOD nMax + 1;
             WriteCh(ch, row, col+in)   (* display purposes only *);
           Release(nTaken);
           Release(inPut);
        END Put;

        PROCEDURE Get(VAR ch:CHAR) ;

        BEGIN
           Claim(inGet);
           Claim(nTaken);
           ch := b[out];
           out := out MOD nMax + 1;
             WriteCh(' ', row, col + out)  (* display purposes *);
           Release(nFree);
           Release(inGet);
        END Get;
   BEGIN
     row:=15; col:=20 (* initialise display part *);
     in:=1; out:=1;
        InitSemaphore(nFree, nMax);
        InitSemaphore(nTaken, 0);
        InitSemaphore(inPut, 1);
        InitSemaphore(inGet, 1);
   END Buffer.
```

Fig. 9.9   IMPLEMENTATION MODULE of buffer using semaphores.

will have to be re-linked in order to pick up the new object code for the changed module.)

The module concept is an important development in the provision of tools for the programming of large systems. It is this idea that makes Modula-2 a strong contender for general programming use.

9.7.2 Low-level facilities

Modula-2 supports a simple set of primitives which have to be encapsulated in a small nucleus coded in the assembly language of the computer on which the system is to run. Access to the primitives is through a module SYSTEM which is known to the compiler. SYSTEM can be thought of as the software bus linking the nucleus to the rest of the software modules (see Figure 9.10). SYSTEM makes available three data types: WORD, ADDRESS, and PROCESS; and six procedures: ADR, SIZE, TSIZE, NEWPROCESS, TRANSFER, and IOTRANSFER.

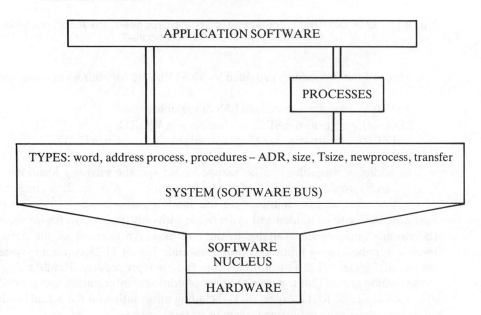

Fig. 9.10 Software structure for Modula-2.

WORD is the data type which specifies a variable which maps onto one unit of the specific computer storage. As such the number of bits in a WORD will vary from implementation to implementation; e.g., on a PDP11 implementation a WORD is 16 bits, on a 68000 it would be 32 bits. ADDRESS corresponds to the definition **TYPE ADDRESS = POINTER TO WORD**; i.e., objects of type ADDRESS are pointers to memory units and can be used to compute the addresses of memory words.

Objects of type PROCESS have associated with them storage for the volatile environment of the particular computer on which Modula-2 is implemented; they make it possible to create process (task) descriptors easily as shown in Figure 9.11.

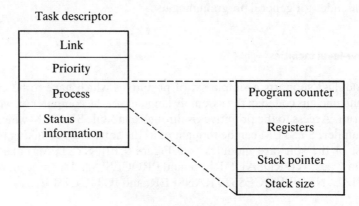

Fig. 9.11 Task descriptor containing machine-dependent PROCESS variable provided by SYSTEM in Modula-2.

Three of the procedures provided by SYSTEM are for address manipulation:

ADR(v)   returns the ADDRESS of variable v
SIZE (v) returns the SIZE of variable v in WORDs
TSIZE(t) returns the SIZE of any variable of type t in WORDs.

In addition, variables can be mapped onto specific memory locations. This facility can be used for writing device-driver modules in Modula-2. A combination of the low-level access facilities and the module concept allows details of the hardware device to be hidden within a module with only the procedures for accessing the module being made available to the end user. An example of the definition module for the analog input and output module for an 11/23 computer system is shown in Figure 9.12. For normal use, the two procedures ReadAnalog and WriteAnalog are all that the user requires. Additional information and procedures are made available to the expert user, including information on the actual hardware addresses (these may vary from system to system).

9.7.3 Concurrent programming: co-routines

Modula-2 provides two simple concurrent programming facilities. At the basic level concurrency on a single processor is provided by co-routines. An implementation-dependent module SYSTEM exports two procedures NEWPROCESS and TRANSFER which are defined as follows:

```
DEFINITION MODULE AnalogIO;
(*--------------------------------------------------------------*)
(*  Sheffield University Control Engineering Department         *)
(*  Analogue Input/Output                                       *)
(*  Version 0, S. White, 2-Jul-85, adapted from ADC and DAC     *)
(*                            modules.                          *)
(*--------------------------------------------------------------*)
FROM SYSTEM IMPORT
          ADDRESS;
EXPORT QUALIFIED
          moduleName, moduleVersion,
          ReadAnalog, WriteAnalog;
CONST
          moduleName = 'AnalogIO';
          moduleVersion = 1;
PROCEDURE ReadAnalog( adcNum: CARDINAL; VAR val: CARDINAL );
PROCEDURE WriteAnalog( dacNum: CARDINAL; val: CARDINAL );
(* Hardware configuration - expert use only *)
TYPE
          AReadProc = PROCEDURE (CARDINAL, VAR CARDINAL );
          AWriteProc = PROCEDURE( CARDINAL, CARDINAL );
PROCEDURE InitAnalog(
               initProc: PROC; readProc: AReadProc;
               writeProc: AWriteProc; doneProc: PROC );
(*------------------ MNCAIO ------------------------------------*)
PROCEDURE InitMNCAIO;
PROCEDURE ReadMNCAD( adcNum: CARDINAL; VAR val: CARDINAL );
PROCEDURE WriteMNCAA( dacNum: CARDINAL; val: CADINAL );
PROCEDURE DoneMNCAIO;
VAR
          MNCADReg: POINTER TO
               RECORD
                    CSR: BITSET;     (* Control/Status *)
                    DBR: CARDINAL;   (* BufferPreset   *)
               END;
          MNCADVec:
               RECORD
                    ConversionComplete: ADDRESS;
                    Error: ADDRESS;
               END;
CONST
     MNCADBaseReg = 171000B;
     MNCADBaseVec = 400B;
VAR
     MNCAAReg: POINTER TO
               RECORD
                    DAC0: CARDINAL;
                    DAC1: CARDINAL;
                    DAC2: CARDINAL;
                    DAC3: CARDINAL;
               END;
CONST
     MNCAABaseReg = 171060B;
END AnalogIO.
```

Fig. 9.12 DEFINITION MODULE for a device handler.

```
PROCEDURE NEWPROCESS(ParameterlessProcedure:PROC;
                    workpaceAddress: ADDRESS;
                    workspaceSize: CARDINAL;
                    VAR coroutine: ADDRESS (* PROCESS *));

PROCEDURE TRANSFER(VAR source, destination : ADDRESS (*PROCESS*));
```

The early definition of SYSTEM included a type PROCESS for co-routines; this type has now been removed and ADDRESS is used in its place though many compilers still use PROCESS.

Any parameterless procedure can be declared as a co-routine. The procedure NEWPROCESS associates with the procedure storage for the process parameters (the process descriptor or task control block – see Chapter 6) and some storage to act as workspace for the process. It is the programmer's responsibility to allocate sufficient workspace. The amount to be allocated depends on the number and size of the variables local to the procedure forming the co-routine and to the procedures which it calls. Failure to allocate sufficient space will usually result in a stack overflow error at run-time.

The variable 'coroutine' is initialized to the address which identifies the newly created co-routine and is used as a parameter in calls to TRANSFER. The transfer of control between co-routines is made using a standard procedure TRANSFER which has two arguments of type ADDRESS (PROCESS). The first is the calling co-routine and the second is the co-routine to which control is to be transferred. An example is given in Figure 9.13. In this example the two parameterless procedures Coroutine1 and Coroutine2 form the two co-routines which pass control to each other so that the message

```
coroutine one    coroutine two
```

is printed out 25 times. At the end Coroutine2 passes control back to the MainProgram.

The fact that the concurrent programming implementation is based on co-routines has led some commentators to state that Modula-2 cannot be used for real-time systems. It cannot be used directly, but it is possible to use the primitive co-routine operations to create a real-time executive as is described in Section 9.7.6 below.

9.7.4 Concurrent programming: Processes

Wirth defined a standard module Processes which provides a higher level mechanism than co-routines for concurrent programming. The module makes no assumption as to how the processes will be implemented; in particular it does not assume that the processes will be implemented on a single processor. If they are so implemented then the underlying mechanism is that of co-routines. The definition

```
MODULE CoroutinesExample ;
(*
Title   : Example of use of coroutines
File    : sb1 coroutine.mod
LastEdit: 20/3/87
Author  : S. Bennett
*)
FROM SYSTEM IMPORT
      ADDRESS, WORD, NEWPROCESS, TRANSFER, ADR, SIZE, PROCESS;

FROM InOut IMPORT
      WriteString, WriteLn;

VAR
      coroutine1Id, coroutine2Id, MainProgram : PROCESS;
      worksp1, worksp2 : ARRAY [1..600] OF WORD;

PROCEDURE Coroutine1;
    BEGIN
      LOOP
        WriteString('coroutine one');
        TRANSFER(coroutine1Id, coroutine2Id);
      END (* loop *);

    END Coroutine1;

PROCEDURE Coroutine2 ;

  VAR
      count : CARDINAL;

    BEGIN
      count:=0;
      LOOP
        WriteString('        coroutine two');
        WriteLn;
        IF count=25 THEN
            TRANSFER(coroutine2Id, MainProgram);
        ELSE
            INC(count);
            TRANSFER(coroutine2Id, coroutine1Id)
        END (* if *);
      END (* loop *);
    END Coroutine2;
BEGIN
  NEWPROCESS(Coroutine1, ADR(worksp1), SIZE(worksp1), coroutine1Id);
  NEWPROCESS(Coroutine2, ADR(worksp2), SIZE(worksp2), coroutine2Id);
  TRANSFER(MainProgram, coroutine1Id)
END CoroutinesExample.
```

Fig. 9.13   Example showing the use of co-routines.

module for Processes is shown below.

```
DEFINITION MODULE Processes ;
    TYPE
          SIGNAL;     (*  opaque type: variables of this type are
                          used to provide synchronization between
                          processes.  The variable must be initialized
                          by a call to Init before use *)
```

```
PROCEDURE StartProcess(P:PROC; workSpaceSize:CARDINAL);
        (* start a new process P is parameterless procedure
        which will form the process, workSpaceSize is the
        number of bytes of storage which will be allocated to
        the process *)

PROCEDURE Send(VAR s: SIGNAL);
        (* If no processes are waiting for s, then SEND
        has no effect.  If some process is waiting for  s
        then  that process is given control and is allowed
        to proceed *)

PROCEDURE WAIT(VAR s: SIGNAL);
        (* The current process waits for the signal s.  If
        at some later time a SEND(s) is is issued by
        another process then this process will return from
        wait. Note if all  processes  are  waiting  the
        program terminates*)

PROCEDURE Awaited(s:SIGNAL):BOOLEAN;
        (* Test to see if process is waiting on s, if one
        or more processes are waiting then TRUE is
        returned*)

PROCEDURE Init(s: SIGNAL);
        (* The  variable  s  is  intitalized,  after
        initialization Awaited(s) returns FALSE*)

END Processes.
```

The procedures and the type SIGNAL exported from the module Processes can be used to solve some of the problems discussed in Chapter 7. The standard module Processes suggested by Wirth and supplied by most systems is not all that versatile: a much more useful and powerful version has been developed at Nottingham University by Henry. It is available both for the DEC 11/23 (RT-11) and for the MSDOS systems – the MSDOS version is available from Logitech Ltd. A description of the system is given in section 9.7.6.

The creation of modules for specific purposes is in keeping with the Modula-2 philosophy. The aim is that the core language should remain fixed and standard and that any extensions required for special purposes should be provided in the form of library modules. It should be noted that all input/output, file handling and other operations are not handled as part of the language, but by standard procedures imported from modules which are assumed to be provided as part of the system.

9.7.5 Interrupts and device-handling

Hardware interrupts can be handled from within a Modula-2 program. A device-handling process can enable the external device and then suspend itself by a call to a procedure IOTRANSFER. This procedure is similar to TRANSFER but has an additional parameter which allows the hardware interrupt belonging to the device to be identified (this is typically to interrupt vector address). When the interrupt occurs control is passed back to the device routine by a return from IOTRANSFER.

The procedure IOTRANSFER has the format

```
IOTRANSFER (VAR interruptHandler : PROCESS;
               interruptedProcess : PROCESS;
               interruptVector : CARDINAL)
```

The action of IOTRANSFER is to save the current status of interruptHandler and to resume execution of interruptedProcess, i.e., to wait for an interrupt. When an interrupt occurs the equivalent of TRANSFER (interruptedProcess, interrupt-Handler) occurs. A skeleton interrupt handler would thus take the form

```
BEGIN
      LOOP
          ...
          IOTRANSFER (interruptHandler, interruptedProcess,
                      interruptVector);
          (* interrupt handler waits at this point for interrupt *)
          ...
      END Loop
END;
```

The interrupt handler code is placed inside the LOOP...END construct, the interrupt handler is initiated by an explicit TRANSFER operation and then waits for an interrupt at the IOTRANSFER statement: at this statement control is returned in the first instance to the initiating task, but when an interrupt has occurred a return will be made to whichever task was interrupted.

Using the facilities provided by SYSTEM a simple foreground/background structure can easily be created to handle real-time control applications.

```
MODULE Main;
FROM SYSTEM IMPORT ADR, SIZE, WORD, PROCESS, NEWPROCESS,
                   TRANSFER, IOTRANSFER;
VAR
      main, operator, control : PROCESS;
PROCEDURE Control;
BEGIN
      LOOP
            IOTRANSFER(control, operator, clockVector);
            ...
            (* control actions go here
                routine should keep track of time as well *)
            ...
      END;
END Control;
PROCEDURE Display;
(* insert the display update code here *)
END Display;
PROCEDURE Keyboard;
(* insert keyboard code here *)
END Keyboard;
PROCEDURE Operator;
BEGIN
   LOOP
     IF time = displayTime THEN Display;
     Keyboard;
   END (* LOOP *);
```

```
END Operator;
BEGIN
    NEWPROCESS (Control, ADR(controlWksp), SIZE(controlWksp),
                control);
    NEWPROCESS (Operator, ADR(operatorWksp), SIZE(operatorWksp),
                operator);
    TRANSFER (main, control);
END Main.
```

It should be noted that in using the low-level facilities directly and simply none of the problems of data-sharing and mutual exclusion discussed in the previous sections have been solved. It is assumed that all the variables required by the controller are stored in common storage and hence are accessible at any time either to the operator task or to the control task. Also, in the above example the control task which is entered on an interrupt can only return to a specific named task and hence there can only be one background task. However, as the next section shows, the low-level facilities provided can be used to create a much more powerful set of real-time multi-tasking support routines.

9.7.6 High-level multi-tasking modules

The low-level facilities provided by SYSTEM enable modules to provide, e.g., signals, semaphores, scheduling etc., to be written in Modula-2. A powerful set of modules for real-time applications have been produced by Henry [Nottingham University]. The general structure of the system is shown in Figure 9.14.

The lowest level module is PROCESSES (a replacement for the Wirth module PROCESSES) which provides the procedures and functions:

Cp   return current process identity.
Disable   disable a given process.
Enable   make a given task runnable.
MinWksp   return the minimum workspace size for task.
NewProcess   create a new task.
PriorityOf   return the priority of a task.
SuspendMe   suspend the calling task.
SuspendUntilInterrupt   suspend the calling task until a specific interrupt occurs.

The relationship between the procedures and the state of the tasks is shown in Figure 9.15. NEWPROCESS is used to inform PROCESSES of the existence of a task, but before it can be run it must be made RUNNABLE by a call to ENABLE. In the call NEWPROCESS a task is allocated a priority (an integer value – the range depending on the implementation) and the runnable task with the highest priority becomes the running task. A running task can call DISABLE which will cause a named runnable task to be changed to EXISTENT (NON-RUNNABLE); if the named task is not runnable the call will be ignored; to make itself non-runnable the running task uses the call SuspendMe. A running task may also suspend itself to wait

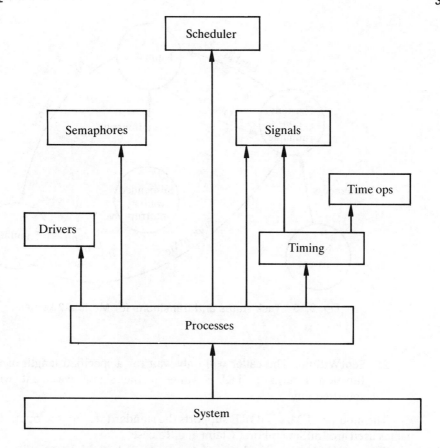

Fig. 9.14 Modula-2 kernel (copyright R. Henry, Nottingham University).

for a hardware interrupt by using the call SuspendUntilInterrupt in which case it will not become runnable until the specified hardware interrupt occurs and is accepted.

For simple applications a high-level module, SCHEDULER, is provided which has two procedures:

1. StartProcess   This is the equivalent of NewProcess and Enable, it makes a task known to Processes and makes it runnable;
2. StopMe   This stops the current task.

The separate modules of SIGNALS, SEMAPHORES and TIMING allow the tasks started by SCHEDULER to synchronize and to run at specified time intervals.

The module SIGNALS provides the standard operations on signals (initialize, wait and send – the names used are InitSignal, SendSignal and AwaitSignal) but in addition two further operations are supported:

1. Awaited   This returns TRUE value if at least one task is waiting for the SendSignal operation.

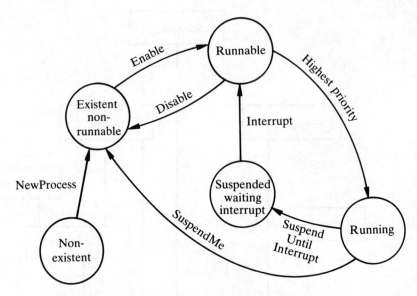

Fig. 9.15 Task states and transitions for Modula-2 kernel.

 2. SentWithin   The caller will only wait for a specified length of time; the function returns a TRUE value if the signal was sent within the specified time.

The module SEMAPHORE supports the standard semaphore operations – the names used are InitSemaphore, Claim and Release.

The module Timing enables users to operate in both *absolute* and *relative* time intervals. Absolute time begins when the Timing module is initialized and one value of absolute time can be said to be earlier or later than another. The difference between two values of absolute time – a time interval – is said to be relative time and one interval can be said to be longer or shorter than another.

Time – both absolute and relative – is measured in units of seconds and ticks. The number of ticks in a second is implementation defined (it depends upon the system clock used) and its value is returned by the function TicksPerSecond. The current time (absolute) can be found using the procedure TellTime. Two procedures are provided to enable tasks to wait for specified times:

 1. DelayFor   The task waits for a specified time interval.
 2. DelayUntil   The task waits until a specified absolute time.

In both cases the calling task is suspended until either the time interval has elapsed or the absolute time is reached; the task is then made runnable – there is no guarantee that the task will run immediately on attaining the specified condition since Processes will choose the task with the highest priority.

In calculating absolute times or time intervals the value of time in seconds and ticks has to be manipulated: to support such operations the module TimeOps provides the following procedures:

1. IncTime – increase a time value by a given interval
2. IncInterval – increase an interval value by a given interval
3. DecTime – decrease a time value by a given interval
4. DecInterval – decrease an interval value by a given interval
5. DiffTimes – subtract second time for the first time
6. DiffIntervals – subtract second interval from first interval
7. CompareTimes – compare first time with second time
8. CompareIntervals – compare first interval with second interval.

A simple example of the use of the RTS kernel facilities is given in Figure 9.16.

```
MODULE TwoTasks ;
(*
Title    : Example of two tasks synchronizing using signals
File     : sb1 twotasks.mod
LastEdit : 20/3/87
Author   : S. Bennett
System   : NUPDS Scheduler and Signals
*)
FROM Scheduler IMPORT
    ProcessId, Priority, StartProcess, StopMe;

FROM Signals IMPORT
    AwaitSignal, SendSignal, InitSignal, Signal;

FROM InOut IMPORT
    WriteString, WriteLn;

  CONST
      priority = 2;
      worksp = 600;
  VAR
    messageSent : Signal;
    count : CARDINAL;
    taskTwoId : ProcessId;

    PROCEDURE TaskTwo ;
    BEGIN
      LOOP
        AwaitSignal(messageSent);
        WriteString(' message received by task two');
        WriteLn
      END (* loop *);

    END TaskTwo;

BEGIN   (* body of program forms task 1 *)
  InitSignal(messageSent);
  StartProcess(TaskTwo, priority, worksp, taskTwoId);
  FOR count:= 1 TO 10 DO
      WriteString('Sending message');
      SendSignal(messageSent)
  END (* for *);

END TwoTasks.
```

Fig. 9.16 Example showing creation of two tasks synchronized using signals.

Two tasks, the task forming the main program and the task formed by procedure TaskTwo are run alternately, synchronized by the use of the signal messageSent. The output is ten lines of 'Sending message' – output by the main task – and 'message received by task two' – output by TaskTwo. TaskTwo is started with a priority of level 2 by the use of the procedure Scheduler.. StartProcess. The main body of the program is automatically run at priority level 0.

Some of the features of the RTS kernel can be explored by using the example program shown in Figure 9.17. Two tasks, TaskA and TaskB, are created; TaskA prints out on the screen a number of rows of the letter A (the number is specified in

```
MODULE RTS3 ;
(*
Title   : Demonstration of resource sharing
File    : sb1: RTS3.mod
LastEdit:
Author  : S. Bennett
System  : NUPDS RTS Kernel
*)
FROM InOut IMPORT
    Write, WriteLn, WriteString;

FROM Semaphores  IMPORT
    InitSemaphore, Semaphore, Claim, Release;

FROM Scheduler IMPORT
      ProcessId, Priority, StartProcess, StopMe;

FROM Timing IMPORT
    DelayFor, DelayUntil, Interval, TellTime, Time, TicksPerSecond;

CONST moduleName = 'RTS3';
      numberOfLines = 5;
VAR
    screen : Semaphore;
    endA, endB : BOOLEAN;

    PROCEDURE TaskA ;
    CONST ch='A';
    VAR
        i,j : CARDINAL;
        delayA :Interval;
    BEGIN
        i:=0; j:=0; delayA.secs:=0; delayA.ticks:=1;
        LOOP
          Claim(screen);      (* omit in versions 1 and 2 *)
          FOR i:=1  TO 79 DO
            Write(ch);
            DelayFor(delayA)   (* omit in version 1 *)
          END (* for *);
          WriteLn;
          Release(screen);    (* omit in versions 1 and 2 *)
          INC(j);
          IF j>numberOfLines THEN
            EXIT
          END (* if *);

        END (* loop *);
        endA := TRUE;
```

Fig. 9.17 Demonstration of resource sharing.

```
            StopMe;
            END TaskA ;
            PROCEDURE TaskB ;
            CONST ch='B';
            VAR
                i,j : CARDINAL;
                delayB : Interval;
            BEGIN
                i:=0; j:=0; delayB.secs:=0; delayB.ticks:=1;
                LOOP
                    Claim(screen);       (* omit in versions 1 and 2 *)
                    FOR i:=1  TO 79 DO
                        Write(ch);
                        DelayFor(delayB);   (* omit in version 1 *)
                    END (* for *);
                    WriteLn;
                    Release(screen);   (* omit in versions 1 and 2 *)
                    INC(j);
                    IF j>numberOfLines THEN
                        EXIT
                    END (* if *);

                END (* loop *);
                endB := TRUE;
                StopMe;
            END TaskB ;
        CONST
            priorityA = 1;
            priorityB = 1;
            wkspSizeA = 1000;
            wkspSizeB = 1000;

        VAR
            taskAId, taskBId : ProcessId;
        BEGIN
          WriteString(moduleName);
          WriteLn;
          endA:=FALSE; endB:=FALSE;
          InitSemaphore(screen,1);
          StartProcess(TaskA, priorityA, wkspSizeA, taskAId);
          StartProcess(TaskB, priorityB, wkspSizeB, taskBId);
          LOOP
            (* Idle process *)
            IF endA AND endB THEN
                EXIT
            END (* if *);
          END (* loop *);
          WriteLn;
          WriteString('Program end');
        END RTS3.
```

Fig. 9.17   (cont.)

the constant numberOfLines). TaskB prints out a number of rows of the letter B. The display which is obtained on the screen depends on the way the tasks are scheduled and whether the tasks are given exclusive access to the screen. By compiling the module in Version 1 form (not using the semaphore and leaving out the DelayUntil calls), the scheduler treats the 'to' tasks as co-routines; hence TaskA which is started first gains control and runs to completion, only then does TaskB run. The result of this is shown in Figure 9.18. The reason for this behavior is that the

```
RTS1
AAAAAAAAAAAAAAAAAAAAAAAAAAAAAAAAAAAAAAAAAAAAAAAAAAAAAAAAAAAAAAAAAAAAAAAAAAAAAA
AAAAAAAAAAAAAAAAAAAAAAAAAAAAAAAAAAAAAAAAAAAAAAAAAAAAAAAAAAAAAAAAAAAAAAAAAAAAAA
AAAAAAAAAAAAAAAAAAAAAAAAAAAAAAAAAAAAAAAAAAAAAAAAAAAAAAAAAAAAAAAAAAAAAAAAAAAAAA
AAAAAAAAAAAAAAAAAAAAAAAAAAAAAAAAAAAAAAAAAAAAAAAAAAAAAAAAAAAAAAAAAAAAAAAAAAAAAA
AAAAAAAAAAAAAAAAAAAAAAAAAAAAAAAAAAAAAAAAAAAAAAAAAAAAAAAAAAAAAAAAAAAAAAAAAAAAAA
AAAAAAAAAAAAAAAAAAAAAAAAAAAAAAAAAAAAAAAAAAAAAAAAAAAAAAAAAAAAAAAAAAAAAAAAAAAAAA
BBBBBBBBBBBBBBBBBBBBBBBBBBBBBBBBBBBBBBBBBBBBBBBBBBBBBBBBBBBBBBBBBBBBBBBBBBBBBB
BBBBBBBBBBBBBBBBBBBBBBBBBBBBBBBBBBBBBBBBBBBBBBBBBBBBBBBBBBBBBBBBBBBBBBBBBBBBBB
BBBBBBBBBBBBBBBBBBBBBBBBBBBBBBBBBBBBBBBBBBBBBBBBBBBBBBBBBBBBBBBBBBBBBBBBBBBBBB
BBBBBBBBBBBBBBBBBBBBBBBBBBBBBBBBBBBBBBBBBBBBBBBBBBBBBBBBBBBBBBBBBBBBBBBBBBBBBB
BBBBBBBBBBBBBBBBBBBBBBBBBBBBBBBBBBBBBBBBBBBBBBBBBBBBBBBBBBBBBBBBBBBBBBBBBBBBBB
BBBBBBBBBBBBBBBBBBBBBBBBBBBBBBBBBBBBBBBBBBBBBBBBBBBBBBBBBBBBBBBBBBBBBBBBBBBBBB
```

Fig. 9.18 Output from Version 1 of resource-sharing program.

scheduler continues to run a task until either a higher priority task wishes to run or until the running task suspends 'for' or ends. Introducing the DelayUntil statements to form Version 2 of the program causes the two tasks to run alternately giving the output shown in Figure 9.19. It is noticeable in Figure 9.19 that some characters are missed as both tasks try to output to the same device.

The introduction of semaphores to provide exclusive access to the screen for one line of output avoids this problem as is shown in Figure 9.20.

```
RTS2
ABABABABABABABABABABABABABABABABABABABABABABABABABABABABABABABABABABABABABABAB
ABABABABABABABABABABABABABABABABABABABABABABABABABABABABABABABABABABABABABAB
A
BABABABABABABABABABABABABABABABABABABABABABABABABABABABABABABABABABABABABA
BABABABABABABABABABABABABABABABABABABABABABABABABABABABABABABABABABABABABAB
A
BABABABABABABABABABABABABABABABABABABABABABABABABABABABABABABABABABABABABA
BABABABABABABABABABABABABABABABABABABABABABABABABABABABABABABABABABABABABAB
A
BABABABABABABABABABABABABABABABABABABABABABABABABABABABABABABABABABABABABA
BABABABABABABABABABABABABABABABABABABABABABABABABABABABABABABABABABABABABAB
A
BABABABABABABABABABABABABABABABABABABABABABABABABABABABABABABABABABABABABA
BABABABABABABABABABABABABABABABABABABABABABABABABABABABABABABABABABABABABAB
A
BABABABABABABABABABABABABABABABABABABABABABABABABABABABABABABABABABABABABA
BABABABABABABABABABABABABABABABABABABABABABABABABABABABABABABABABABABABABAB
```

Fig. 9.19 Output from Version 2 of resource-sharing program.

```
RTS3
AAAAAAAAAAAAAAAAAAAAAAAAAAAAAAAAAAAAAAAAAAAAAAAAAAAAAAAAAAAAAAAAAAAAAAAAAAAAAA
BBBBBBBBBBBBBBBBBBBBBBBBBBBBBBBBBBBBBBBBBBBBBBBBBBBBBBBBBBBBBBBBBBBBBBBBBBBBBB
AAAAAAAAAAAAAAAAAAAAAAAAAAAAAAAAAAAAAAAAAAAAAAAAAAAAAAAAAAAAAAAAAAAAAAAAAAAAAA
BBBBBBBBBBBBBBBBBBBBBBBBBBBBBBBBBBBBBBBBBBBBBBBBBBBBBBBBBBBBBBBBBBBBBBBBBBBBBB
AAAAAAAAAAAAAAAAAAAAAAAAAAAAAAAAAAAAAAAAAAAAAAAAAAAAAAAAAAAAAAAAAAAAAAAAAAAAAA
BBBBBBBBBBBBBBBBBBBBBBBBBBBBBBBBBBBBBBBBBBBBBBBBBBBBBBBBBBBBBBBBBBBBBBBBBBBBBB
AAAAAAAAAAAAAAAAAAAAAAAAAAAAAAAAAAAAAAAAAAAAAAAAAAAAAAAAAAAAAAAAAAAAAAAAAAAAAA
BBBBBBBBBBBBBBBBBBBBBBBBBBBBBBBBBBBBBBBBBBBBBBBBBBBBBBBBBBBBBBBBBBBBBBBBBBBBBB
AAAAAAAAAAAAAAAAAAAAAAAAAAAAAAAAAAAAAAAAAAAAAAAAAAAAAAAAAAAAAAAAAAAAAAAAAAAAAA
BBBBBBBBBBBBBBBBBBBBBBBBBBBBBBBBBBBBBBBBBBBBBBBBBBBBBBBBBBBBBBBBBBBBBBBBBBBBBB
AAAAAAAAAAAAAAAAAAAAAAAAAAAAAAAAAAAAAAAAAAAAAAAAAAAAAAAAAAAAAAAAAAAAAAAAAAAAAA
BBBBBBBBBBBBBBBBBBBBBBBBBBBBBBBBBBBBBBBBBBBBBBBBBBBBBBBBBBBBBBBBBBBBBBBBBBBBBB
```

Fig. 9.20 Output from Version 3 of resource-sharing program.

Modula-2 is an excellent language which meets most of the requirements outlined in the previous chapter. It has attracted considerable interest and offers a simpler, but still powerful, alternative to Ada. Widespread adoption will be dependent on the range and quality of the support libraries available and agreement on standard I/O libraries and standard real-time support modules.

9.8 ADA

The language Ada arose out of a survey carried out by the US Department of Defense in the early 1970s when they discovered that the systems they were using involved over 300 different programming languages. As a result of this survey the Department decided that a high-level language must be developed and that after a certain date all operational software would have to be written in the new language. Between 1976 and 1979 extensive work took place on drawing up specifications and on evaluation exercises. The final design stage took the form of a competition and out of this competition Ada was selected.

The intention in developing the new language was that it should conform to modern ideas about software engineering, particularly in relation to the construction of large-scale real-time systems. There was in the specification considerable emphasis on 'readability' with the aim of reducing documentation and maintenance costs. In order to avoid the development of 'dialects' the name Ada has been registered as a trade mark and any compiler which is called Ada must be validated by the Department or another authorized centre.

As has been already noted Ada is an even more strongly typed language than Pascal or Modula-2 in that derived types, for example, are treated as being incompatible with their parent types. All mixed expressions require explicit type conversion.

Careful attention has been paid to the nature of compilation units and the ability to carry out separate, rather than independent, compilation. The solution adopted, the PACKAGE, is similar to that used in Modula-2. A package is divided into two parts: the specification and the body. The specification is public and contains the definition of the interface between the package and the client program, i.e., it defines types, data objects, procedures and functions. The body is private to the package implementor. The mechanism is a little more powerful than the Modula-2 equivalent, in that, e.g., a private type can be declared. This allows the user to declare variables of the particular type, but the user does not know any details of the type. Therefore, just as the code can be changed without affecting the user, the type implementation can also be changed.

The package mechanism (and sub-programs) have been extended further through the concept of the creation of generics. This mechanism allows the same code to be applied to different types of data. For example a sort routine can be written which can then be instantiated to operate on, in one case, real variables and, in another case, integer variables.

Ada supports an elegant multi-tasking mechanism which solves simply all the mutual exclusion and synchronization problems discussed in Chapter 7; it also provides an exception handling mechanism.

9.9 APPLICATION-ORIENTED SOFTWARE

A large number of software packages and languages have been developed with the intention of providing a means by which the end user can easily write or modify the software for a particular problem. The major reason for wanting the end user, rather than a specialist programmer, to write the software is to avoid the communication problem. A large proportion of 'errors' in a system arise from misunderstanding of the operation or structure of the plant by the programmer. It should not be assumed that this is always the programmer's fault; often the engineers and managers responsible for the plant do not communicate their requirements clearly and precisely. The misunderstandings can largely be avoided if the engineers responsible for the plant can themselves write the software. A second reason is economic: high quality, specialist computer programmers are in short supply and hence are expensive to employ.

The engineers are not, however, expected to be specialists in computer programming and hence they must be provided with simple programming tools which reflect the particular application. Because for a given type of application, e.g. process control, the range of facilities required is small and predictable, it is not too difficult to devise special application software.

Three main approaches are used:

1. table-driven;
2. block-structured; and
3. specialized languages;

and these are considered in more detail in the following sections.

9.9.1 Table-driven

The table-driven approach arose out of systems programmed in assembly code. It was soon realized that many code segments were used again and again in different applications as well as in the particular application. For example, the code segment used for PID control need only be written once if it contains only pure code and all references to parameters and data are made indirectly.

A simple table-driven system is illustrated in Figure 9.21; as well as allowing the control program to communicate with the data and parameter table, provision is also made for the operator to obtain information from the table (and in some instances to change values in the table). In addition some systems allow the user to write application programs in a normal computer language (usually FORTRAN or

Application-oriented Software

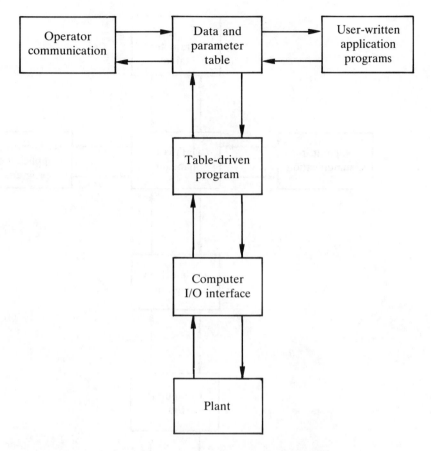

Fig. 9.21 Simple table-driven system.

BASIC) which can interact with the table-driven software. The actual table-driven program is supplied as part of the system and cannot be modified by the user. It will typically have been written in assembly language and will have been written with an emphasis on efficiency and security.

An alternative approach is shown in Figure 9.22, in which a database manager program is inserted between the data table and all users; access to the data table is now controlled by the database manager. The use of a database manager to control access places additional overheads on the system and can slow it down. It has the advantage of improving the security of the system in that checks can be built into it to, e.g., limit the items which can be changed from the operator's console, or to restrict the access of the user-written application programs to certain areas of the data table. Typically a database manager program would provide, by means of a password or a key-operated switch on the console, different access rights to the operator and the plant engineer.

A crucial factor in the usefulness of table-driven software is the method of

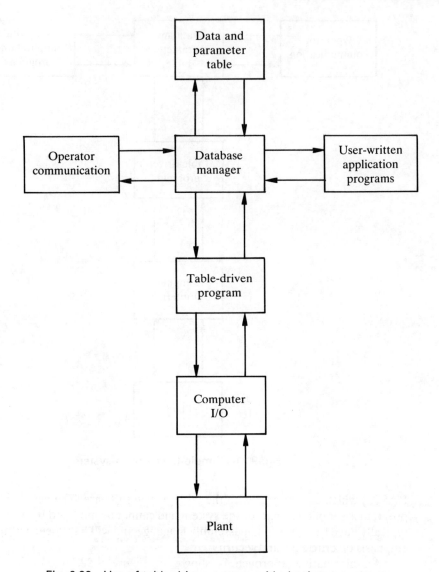

Fig. 9.22 Use of table-driven system with database manager.

setting up and modifying the tables. There are three main methods:

1. direct entry into the data tables;
2. use of language DATA statements; and
3. filling in of forms.

In the simplest systems the data has to be entered into specified memory locations, but this method is rarely used nowadays. It is more normal for the entry program to allow names to be used for the locations; thus, for example, in a system with eight analog inputs the conversion values for each channel may be set by entering from a keyboard statements such as:

```
ANCONV(1) = 950.2
ANCONV(6) = 0.328
```

which would set the conversion factors for the signals coming in on channels 1 and 6. On some systems it is possible to specify signal names rather than use the table index numbers. For example,

```
ANINP(1) = FEEDFL
ANINP(2) = FEEDTP
```

would specify that the signals on analog inputs 1 and 2 are to be known as FEEDFL (representing, say, feedwater flowrate) and FEEDTP (representing feedwater temperature). Once such names have been declared they can be used elsewhere to reference the particular signals. This is a feature which is useful when user-written application software is used, say, to produce specialized displays.

The use of language statements to set up the data tables is useful in large systems for initially setting up the system. It is normally supplemented by allowing change to be made from the operator keyboard, either of all entries or of selected entries. It has the advantage that meaningful names can be given to the entries at initial setting-up time; subsequent modifications are then made using the plant names rather than the table entry index.

The most commonly used approach is to provide a form into which data relating to system is entered. In the early systems data was transcribed on to punched cards and then read into the system; now the form is presented on the screen and the data entered directly. Usually some type of 'forms' processor is provided which checks data for consistency as it is being entered and prompts for any missing data (Figure 9.23).

Table-driven software is very simple to use; it is, however, restrictive in that maximum numbers for each type of control (loop, input, and output) have to be inserted when the system is configured and these cannot be changed by the user. Because of this a similar, but more flexible, approach based on the use of function blocks has been developed.

9.9.2 Block-structured software

The software supplied in a block-structured system consists of a library of function blocks (scanner routines, PID control, output routines, arithmetic functions, scaling blocks, alarm routines, and display routines), a range of supervisory programs and programs for manipulation of the block functions. The engineer programs a control scheme by connecting together the various function blocks which he or she requires and entering the parameters for each block. This is typically done using a VDU with a graphical representation of the block connections. An example of the information which has to be supplied is shown in Figure 9.24.

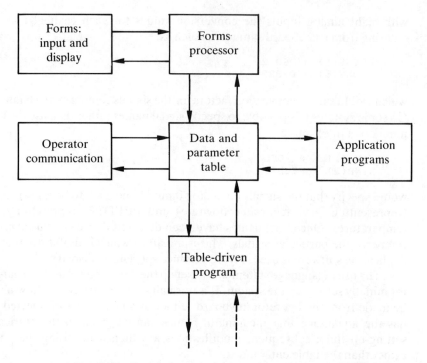

Fig. 9.23 Forms processor for table-driven system.

The block structured approach is used in a wide range of systems, from large process control systems with several hundred loops and multiple operator display stations to simple programmable controllers used for sequence control. The range of controllers available is shown in Figure 9.25.

9.9.3 Application languages

Application languages range from simple interpreters which allow for interaction with table-driven or functional block systems through to complex high-level languages which have to be compiled. The major feature of such languages is that they provide a syntax which reflects the nature of the application. In the following section a large complex application language is described in outline.

9.10 CUTLASS

CUTLASS is high-level language which is oriented towards use by the engineer rather than the professional programmer. It has been developed by the UK Central Electricity Generating Board with the aim of enabling engineering staff to develop,

```
If the input is from another block, enter the block number.            5)   _ _ _ _ _
         or
If the input is from the PCM, enter the register number. (0-59)       5)   _ _
         or
If the input is from the Analog Input Module, enter the following:
    Analog Input Module Type (1=contact, 2=fixed gain, 3=programmable gain,
                              4=Interspec)                             5)   _

If type 1, enter the following information:
    Multiplexer address. (0-1023)                                      5)   _ _ _ _
    Gain code. (0=1V, 2=50MV, 3=10MV)                                        _

If type 2, enter the following information:
    Nest address. (0-15)                                               5)   _ _
    Card address. (0-13)                                                    _ _
    Point address. (0-7)                                                    _
    Gain code. (0=X₁, 1=X₂, 2=X₃, 3=X₄)                                     _

    (Note: X₁-X₄ are defined at SYSGEN time)

If type 3, enter the following information:
    Nest address. (0-15)                                               5)   _ _
    Card address. (0-13)                                                    _ _
    Point address. (0-7)                                                    _
    Gain code. (3=1V, 4=500MV, 5=200MV, 6=100MV, 7=50MV, 8=20MV,
                9=10MV)                                                     _
    Bandwidth. (0=1KH, 1=3KH, 2=10KH, 3=100KH)                              _

If type 4, enter the following information:
    ISCM number. (1-3)                                                 5)   _
    CCM number. (1-16)                                                      _ _
    Type of input. (M=Measurement, S=Setpoint, 0=Output)                    _
    Point number. (1-16)                                                    _ _
```

```
Range of the input in engineering units:
    Lowest value. (-32767. to +32767.)                                 6)   _ _ _ _ _ _
    Highest value. (-32767. to +99999.)                                     _ _ _ _ _ _
    Units. (As specified by user at System Generation.)                     _ _ _ _
```

```
Signal conditioning index. (0-7)                                       7)   _
Thermocouple type if thermocouple input is through the Analog
Input Module. (J, K, T, R) [Otherwise enter N.]                             _
Linearization polynomial index. (0-511) [For signal conditioning indexes
0 or 5 only Enter 5 only if input is from PCM.]                             _ _ _
```

```
Is digital integration required? (Y or N)                              8)   _
    If Y, enter integration multiplier K1, (1-32767)                        _ _ _ _ _
    and integration divisor K2. (1-32767)                                   _ _ _ _ _
    If N, enter the smoothing index. (0-63)                                 _ _
```

```
Operator Console Number (1, 2, or 3)                                   11)  _
```

```
Process unit number (1-127;0 = none)                                   12)  _ _ _
```

```
Block description for alarm messages. Leading and imbedded blanks will be   13)  _ _ _ _ _ _ _
included.                                                                        _ _ _ _ _ _ _
                                                                                 _ _ _ _ _ _ _
                                                                                 _ _ _ _ _ _ _
```

```
Is a supervisory program called when an alarm occurs? (Y or N)         15)  _
    If Y, enter program call number. (0-2047)                               _ _ _ _
```

```
Should this block inhibit the passing of initialization requests? (Y or N)  18)  _
```

```
If an input fails, should this block continue control using the last good   19)  _
value? (Y or N)
```

Fig. 9.24 Typical layout sheet for forms entry (reproduced with permission from Mellichamp, *Real-time Computing*, 1983, Van Nostrand Reinhold).

Software functions:	Basic	Advanced	Process
	Boolean	Block transfer	Signalling
	Timers	Jump	Monitoring
	Counters	File	PID-control
	Data move	Shift registers	Communication
	Comparison	Sequencers	Logging
	Arithmetic	Floating point	Display
Hardware functions:	Small PCs		Large PCs
Inputs	16		4096
Outputs	16		4096
Timers	8		256
Counters	8		256
User program	2 K		48 K
Cycle time (per 1 K)	100 ms		1 ms

Fig. 9.25 Block functions in programmable controllers.

modify and maintain application software independently of professional software support staff.

The major requirements which CUTLASS had to meet are:

1. It should be suitable for a wide range of applications within power stations. All applications packages should operate within the same general framework so that future developments can easily be incorporated without rendering obsolete previous work. To achieve these aims the language was developed in the form of a number of compatible subsets which cover the following functions:
   (a) modulating control;
   (b) data logging;
   (c) data analysis;
   (d) sequence control;
   (e) alarm handling;
   (f) visual display; and
   (g) history recording.

   These subsets operate within a framework which provides
   (a) a real-time executive – TOPSY;
   (b) communications network management;
   (c) support facilities; and
   (d) integrated I/O and file handling.

2. The software should have a high degree of independence from any particular computer type. This is a particularly important requirement for software which is expected to have a long lifetime – twenty to thirty years – during which period the actual control computers may have to be replaced and it is important that this can be done without having to either use

obsolete technology or incur high program modification costs. In order to achieve this the CUTLASS language has been written in CORAL.

3. The software should be simple and safe to use so that engineers based in the power station can produce and modify programs. This has been achieved by providing within the language an extensive range of sub-routines and by hiding the detailed operational and security features from the user.

9.10.1 General features of CUTLASS

The basic unit of a CUTLASS program is a SCHEME which is an independently compilable unit. A SCHEME is defined as

```
<subset type> SCHEME <name>
    GLOBAL   <data>
    COMMOM   <data>
    TASK <qualifying data>
    ...
    TASK <qualifying data>
ENDSCHEME
```

A scheme may contain any number of tasks and a program may contain any number of schemes. The schemes may run on different computers in a distributed network. The software is developed on a host machine and downloaded to the target machine(s); a typical system is shown in Figure 9.26. During the operation of the overall system the host may remain connected to the target systems.

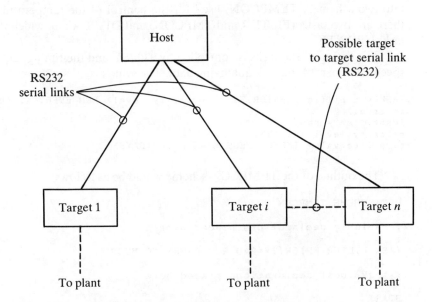

Fig. 9.26 CUTLASS host-target configuration.

In addition to the schemes generated by the programmer, support software – including the TOPSY executive – is loaded into the target machine. The schemes can be enabled and disabled by the user from a keyboard connected to the target machine or from the host machine or from within a supervisory task.

As an example of the division of the software for an application into schemes and tasks, consider the hot-air blower system described in Chapter 1. For this system it is required that the temperature measurement be filtered by taking a running average over four samples at 10 ms intervals. The actual control is to be a PID controller running at 40 ms intervals. The display of the input, output and error is to be updated at 5 s intervals.

```
1. TEMPERATURE CONTROL SCHEME

   1A TASK FILTER RUN EVERY 10 ms
      Reads temperature from heater and computes the running
      average.
   1B TASK CONTROL RUN EVERY 40 ms
      Uses the average temperature obtained from filter and
      computes, using a PID control algorithm the output for
      the heater.

2. DISPLAY SCHEME

   2A TASK DISPLAY UPDATE RUN EVERY 5 s
      Update the display with the values of temperature, error
      and heater output.
```

Fig. 9.27   Scheme and task outline.

A possible arrangement is shown in Figure 9.27 in which the program is split into two schemes: TEMPCON, used for the control of the temperature in which there are two tasks (FILTER and TMPCON) and DISPLAY in which there is one task, UPDATE.

The timing of the tasks is controlled by TOPSY and the timing requirement is specified as part of the task qualifying data. The syntax is

```
TASK <identifier> PRIORITY = <priority level> RUN EVERY <integer> <time
interval>
identifier = name
priority level = 1..250
time interval = {MSECS|SECS|MINS|HOURS|DAYS}
```

The outline of the TEMPCON scheme would be as follows

```
DDC SCHEME TEMPCON
;
; variable declarations placed here
;
TASK FILTER PRIORITY=240 RUN EVERY 10 MECS
;
; task local declarations placed here
;
START
;
```

```
; task body
;
ENDTASK
;
TASK TMPCON PRIORITY=225 RUN EVERY 40 MSECS
;
; task declarations
;
START
;
; task body
;
ENDTASK
ENDSCHEME
```

As is seen from the above statements, the TOPSY executive supports CYCLIC or CLOCK-based tasks; it also supports a DELAY timing function, but the DELAY function can be used only within the GEN subset, and then only for tasks which are non-repetitive.

9.10.2 Data typing and bad data

As in Pascal, CUTLASS requires all variables to be associated with a data type when they are declared. The types supported are: logical, integer, real, text, string, array and record.

An additional feature which is used in CUTLASS is that variables can be flagged as good or bad. The concept of 'bad' data is useful in an environment in which a large number of values are derived from plant measurements. Instruments on the plant can become faulty and supply data which is incorrect. It is often easy to detect when an instrument is supplying false information as when, for example, typical plant transducers supply signals in the range 4–20 mA; if the signal falls below 4 mA it is an indication that the transducer is faulty. The difficulty arises in transmitting through the system an indication that the reading from that particular instrument is faulty.

The ultra-safe approach would be to put the whole scheme into manual mode and allow the operator to take over until the instrument is repaired or replaced. In many circumstances such extreme action may be unnecessary: the instrument may only be supplying an operator display, or there may be an alternative measurement available.

The approach adopted is to mark the data value as bad and allow the bad data to propagate through the system, action, if necessary, being taken elsewhere in the system. The rules used in the propagation of bad data produce a result which is marked as bad if any of the operands in an arithmetic operation are bad. For example

```
A := B + C
E := D/C
```

would result in E being marked as bad if B were bad. The rules for propagation through logic operations are slightly more complex in that a logic variable which is bad is treated as a 'don't know' condition. For example:

```
MODEA := (TEMP > 30.0) OR FANRUN
```

would produce a valid result if TEMP was bad and FANRUN was TRUE, but a bad value if FANRUN was FALSE. An advantage of the bad data flagging is that large sections of the code can be written on the assumption that the data is good: if it is not good then the fact that it is bad will be automatically propagated through the system to the point at which action must be taken.

9.10.3 Language sub-sets

The language is divided into four subsets which share some common features but which perform essentially different functions. A scheme may use only instructions from one subset and the subset being used is declared as part of the scheme heading. The subsets are:

1. GEN A general purpose subset which is used to support text input and output; a scheme which uses only the features common to all subsets is also referred to as a GEN scheme.
2. DDC This subset provides the set of instructions used for direct digital control algorithms.
3. SEQ A subset which supports the construction of sequential control algorithms; there is some restriction on the use of the common language features within a SEQ scheme.
4. VDU A subset which provides graphical and text support for a range of VDUs including both color and monochrome devices.

Some indication of the power of the DDC subset can be obtained from the list of instructions supported which is shown in Figure 9.28.

9.10.4 Scope and visibility

The CUTLASS system has very simple scope and visibility rules. Objects declared within a task are local to that task: the scope extends throughout the task and the object is visible within the body of the task. The object is not visible within a sub-routine called by the task.

Variables may be shared between tasks by declaring them in a COMMON block in the scheme declarations. The compiler restricts variables declared within the COMMON block to the sharing of information between tasks and prevents their use as local variables as well by enforcing the following rules:

1. Only one task may write to a COMMON variable. For this purpose arrays are treated as indivisible objects so that if a task writes to an individual element of an array then all the elements become write only for that task.
2. Tasks are not allowed read access to variables to which they write.

```
FUNCTION                OUTPUT
Non-History Dependent Functions

AVE         average of all good inputs
EAVE        average of all inputs, if any input is bad then output is bad
MIN         minimum of good inputs
AMIN        absolute minimum of good inputs
EMIN        minimum of inputs, if any input is bad then output is bad
EAMIN       absolute minimum of inputs, if any is bad then output is bad
MAX         maximum of good inputs
AMAX        absolute maximum of good inputs
EMAX        maximum of inputs, if any is bad then output is bad
EAMAX       absolute maximum of inputs, if any bad then output is bad
INHIBIT     depends on condition of inhibit raise and lower flags
INCS        converts integer value to incremental units
LIMIT       limit range of real variable
DBAND       deadband function

History Dependent Functions

BUCKET      counts increments supplied and used
FLIPFLOP    logical flipflop triggered at twice the task repeat interval
CHANGE      logic function $A_n = A_n$ EOR $A_{n-1}$
RISING      logic function $A_n = A_n$ AND NOT $(A_{n-1})$
FALLING     logic function $A_n = $ NOT$(A_n)$ AND $A_{n-1}$
ACC         sets accumulator to an initial condition
ACCLIMIT    sets hard limits on an accumulator

History and Time Dependent Functions

INT         approximate integrator
DELTA       approximate differentiator
FIRST       first order filter
INCPID      incremental PID control algorithm, includes roll-off filter
PID         absolute PID algorithm including roll-off filter
ZFORM       controller expressed as a polynomial in z
RAMP        limits rate of change of a variable.
```

Fig. 9.28   CUTLASS – functions available in DDC subset.

Communication between tasks in different schemes is by means of GLOBAL variables. These variables can be created only by use of a special utility which is run by a privileged user. GLOBAL variables are owned by the user who created them and may be removed only by the owner. The same access rules as are applied to GLOBALs with the added restriction that only a task belonging to the owner of a GLOBAL variable may write to that variable.

The rules are relaxed for tasks which belong to users with privileged status such that

1. GLOBALs may be declared as private to the user.
2. GLOBALs may be written to by more than one task (subject to the ownership rule).
3. Tasks may have both read and write access to the variable.

9.10.5 Summary

CUTLASS represents an attempt to resolve many of the problems which arise in real-time systems. It does not purport to be a general purpose language and hence the solutions which are adopted are not always particularly elegant. The emphasis in the system is on enabling inexperienced computer users to write reliable, secure, programs. In this the language is successful.

However, the overall system is complex and setting up and maintaining it requires the support of experienced computer staff. This does not detract from one of the aims of the system, which was to allow engineers to write their own application programs: this has been achieved. It would be difficult, however, to operate the CUTLASS system in an organization which did not have expert computer staff.

Although the language syntax is an improvement on CORAL 66, readability is not high in that identifiers are restricted to nine characters and must be in upper case. A further restriction is that user sub-routine libraries cannot be developed; all user (as opposed to system) sub-routines must be included within the TASK declaration area in source code form. The language supports constants in a useful form, variables can be preset and such variables are then treated as read only variables.

A flavor of the style of the language can be obtained from the scheme listed in Figure 9.29.

9.11 CHOICE OF PROGRAMMING LANGUAGE

In engineering, change is usually gradual: it is only when progress along a particular path becomes completely blocked that major changes occur. There is thus a reluctance to accept new languages however overwhelming the case for acceptance appears to be to the computer scientist. A major reason for not changing is the accumulated experience and knowledge of the existing language: this accumulation of knowledge and experience leads to the production of safe, reliable and efficient programs. The change to a new language involves a learning period which is costly in that it requires the retraining of staff, overcoming of customer resistance, and close monitoring of the first projects using the new language. It is therefore to be expected that for many years engineers will continue to use the older languages: assemblers, BASIC, and FORTRAN.

Another reason for the reluctance to change is the availability of tools to support a particular language: context editors, cross-reference generators, code control systems, execution flow analyzers, debugging systems, etc. The quality of these tools may have a much more significant impact on the development costs and the final quality of the software than the actual language itself.

The development and maintenance costs can be considerably reduced if a language is well supported by a range of development tools. Development tools would include all, or some, of the following

- editor(s);
- library manager;

```
[002,00
DDC SCHEME AMILLFEED
        ;The purpose of this scheme is to maintain the air/fuel ratio
        ;within the mill to a desired and safe operating condition.
        ;This is achieved by the adjustment of the actuator on the
        ;coal feeder which varies the fuel quantity input to the mill.
        ;The primary parameters for the control of the air/fuel ratio
        ;are the mill and primary air differential pressures.
        ;Version 1,00  Written for Unit 2  F S Peach 4 Oct 83
        ;Version 2,00  Stripped down for publication  M L Bransby 9 Dec 83

        ;Declare global data
GLOBAL
        ;Globals preset by the Monitor and used by this scheme.
REAL     RGAINS[GN1]
INTEGER  IGAINS[GN2]
        ;Communication globals - to primary air fan scheme.
LOGIC    FAUTOA
        ;Declare common data
COMMON
REAL     MDY                     ;Filtered mill differential pressure
REAL     PADY                    ;Filtered primary air differential pressure
LOGIC    M                       ;Flip flop flag
INTEGER  N                       ;Increments to be sent
        ;Declare IO information as common presets
INTEGER  A1:=0,CH1:=2            ;Mill diff
INTEGER  A2:=0,CH2:=3            ;PA diff
INTEGER  A3:=4,CH3:=3            ;Mill motor current
INTEGER  A4:=20                  ;Feeder actuator position
INTEGER  DI1:=50,B1:=13          ;Auto-Man button
INTEGER  DO1:=64,B3:=9,B4:=7,B5:=8  ;Auto status,lower,raise
;**********************************************************************
TASK FEED PRIORITY=210 RUN EVERY 5 SECS
        ;This task performs the main control functions

        ;Declare local data
        ;Real parameters from global array RGAINS[ ]
REAL     K                       ;Gain (%/mbar of mill diff)
REAL     TI                      ;Integral time (secs)
REAL     TD                      ;Derivative time (secs)
REAL     TF                      ;Roll off time (secs)
REAL     VMIN                    ;Min actuator position (%)
REAL     VMAX                    ;Max actuator position (%)
REAL     SC                      ;Scale (% actuator movement/pulse)
REAL     RATIO                   ;Desired mill diff/PA diff ratio
REAL     MMCOP                   ;Mill motor current running (amps)
        ;Real inputs from Media
REAL     MD                      ;Scaled mill diff
REAL     PAD                     ;Scaled PA diff
REAL     MMCX,MMC                ;Raw,scaled mill motor amps
REAL     VX,V                    ;Raw,scaled actuator position
        ;Derived real variables
REAL     ERROR                   ;Pressure error from desired
REAL     DX0                     ;Incpid block output
REAL     REM                     ;Demanded output not yet sent
REAL     DX                      ;Incpid output + any remainder
REAL     VDOWN,VUP               ;Actuator pos distance from bottom,top
REAL     DX1                     ;Limit block output
REAL     SENT                    ;Output being sent-not used
```

Fig. 9.29 Example of a CUTLASS DDC scheme (reproduced with permission from Bennett and Linkens, *Real-time Computer Control*, 1984, Peter Peregrinus).

```
            ;Integers from array IGAINS[ ]
    INTEGER NMIN,NMAX            ;Min,max no of pulses
            ;Derived integers
    INTEGER N2                   ;Pulses to be output
            ;Logics from digital inputs
    LOGIC   AMB,AMBX             ;Auto manual button
            ;Derived logic variables
    LOGIC   GDSCH                ;Good signals
    LOGIC   MMRUN                ;Mill motor running
    LOGIC   OK                   ;Auto available
    LOGIC   AUTO                 ;Auto available and selected
    LOGIC   HI,LO                ;Limit block outputs-not used
;-----------------------------------------------------------------
START
MXANIN'MED CARD A3 CH3,MMCX      ;Read mill motor current
ANIN'MED A4,VX                   ;Read actuator position
DIGIN'MED CARD DI1 B1, AMBX      ;Read auto manual button
            ;Scale the analogue inputs into engineering units
MD:=(MDY-0.1)*70.0/0.8           ;Scale mill diff 0-70 mbar
PAD:=(PADY-0.1)*7.0/0.8          ;Scale PA diff 0-7 mbar
MMC:=(MMCX-0.1)*120.0/0.8        ;Scale mill amps 0-120 amps
V:=(VX-0.1)*100.0/0.8            ;Scale act. pos. 0-100%
            ;Invert digital inputs so that closed Media contacts show true
AMB:=NOT(AMBX)
            ;Preset control parameters set under Monitor
K:=RGAINS[1]
TI:=RGAINS[2]
TD:=RGAINS[3]
TF:=RGAINS[4]
VMIN:=RGAINS[5]
VMAX:=RGAINS[6]
SC:=RGAINS[7]
RATIO:=RGAINS[8]
MMCOP:=RGAINS[9]
NMIN:=IGAINS[1]
NMAX:=IGAINS[2]
            ;For the scheme to be able to operate in auto it is necessary
            ;that all the inputs from plant and all the globals set using
            ;the Monitor are good. It is possible to check each variable
            ;explicitly, but this is heavy. The alternative approach is to
            ;note that the final control output from this task, N2, is only
            ;good if all the variables used to generate it are good. This
            ;may therefore used to check all the inputs in one go. This
            ;works whatever the mode of the scheme.
GDSCH:=GOODINT(N2)
            ;The mill motor must be running before auto available is displayed
MMRUN:=MMC>MMCOP
            ;Calc the auto available flag for display on the desk A/M station
OK:=GDSCH AND MMRUN
            ;The scheme will be put into Auto-Normal mode when auto is available
            ;and auto is selected on the desk.
AUTO:=OK AND AMB
AUTOMAN AUTO
ERROR:=RATIO*PAD-MD                      ;Compute mill diff. error
            ;The mbar pressure error is the input to the controller. the
            ;output from the controller is % movement per mbar error.
DX0:=INCPID(ERROR,K,TI,TD,TF)            ;Feeder controller
            ;Summate the output change demanded this tick with that demanded
            ;in previous ticks, but not sent due to the actuator rate limit,
            ;quantisation or minimum movement.
```

Fig. 9.29 *(cont.)*

```
        DX:=DX0+REM                             ;Total % movement wanted this tick
                    ;The summator output is passed to a look-ahead limit.Here the
                    ;demanded movement is compared with the movement required to
                    ;reach an actuator max or min limit so that these are not exceeded.
        VDOWN:=VMIN-V                           ;Distance to bottom
        VUP:=VMAX-V                             ;Distance to top
        LIMIT OUTPUT DX1,LO,HI INPUT DX,VDOWN,VUP
                    ;The incs block receives the limit block output of demanded
                    ;movement and converts it into a number of 320ms drive pulses.
        INCS OUTPUT N2,SENT,REM INPUT DX1,SC,NMIN,NMAX
        N:=N2                                   ;Write pulses to common
                    ;Generate a flip flop flag to indicate to the FEEDPULSE task
                    ;when this task has run.
        FLIPFLOP M
                    ;The master pressure controller is not allowed to vary the primary
                    ;air flow demand unless this scheme is on auto as this could lead
                    ;to an incorrect air/fuel ratio.
        FAUTOA:=AUTO                            ;Write scheme status to global
                    ;Set the digital output to the desk auto-manual station.
                    ;           false,true,bad
        DIGOUT'MED OK CARD DO1,B3 CLEAR,SET,CLEAR        ;Output to AM station
        ENDTASK
        ;****************************************************************
        TASK FILTER PRIORITY=220 RUN EVERY 1 SECS
                    ;This task reads the mill and PA diff signals five times faster
                    ;than used in the control task and filters them to remove noise.

                    ;Declare local data
        REAL    MDX                             ;Raw mill diff
        REAL    PADX                            ;Raw PA diff
        REAL    TMD                             ;Mill diff filter time const (secs)
        REAL    TPAD                            ;PA diff filter time const (secs)
        ;----------------------------------------------------------------
        START
        TMD:=RGAINS[10]
        TPAD:=RGAINS[11]
        MXANIN'MED CARD A1 CH1,MDX              ;Read mill diff
        MXANIN'MED CARD A2 CH2,PADX             ;Read PA diff
        MDY:=FIRST(MDX,TMD)
        PADY:=FIRST(PADX,TPAD)
        ENDTASK
        ;****************************************************************
        TASK FEEDPOS PRIORITY=230 RUN EVERY 320 MSECS
                    ;Due to restriction on multi-write to common we cannot pulse N.So
                    ;put N1 into a buffer and reset this to N each time FEED task runs.

                    ;Declare local data
        LOGIC   M0                              ;Change flag
        INTEGER N1                              ;Pulses to be output
        ;----------------------------------------------------------------
        START
        M0:=CHANGE(M)
        IF M0 TRUE THEN
        N1:=N
        ENDIF
        PULSE'MED N1 CARD DO1 LOWER B4 RAISE B5
        ENDTASK
        ENDSCHEME
```

Fig. 9.29 (cont.)

- linker/loader;
- debugger;
- version control;
- database manager;
- pretty printer; and
- cross-reference generator.

The editor would normally be a screen-based editor with extensive facilities for search/replace, block manipulation (cut and paste), file handling, etc. Many integrated language systems have editors which are context-sensitive, which have templates for standard language constructs and syntax checkers for the code entered. In such systems it is normal to be able to run the compiler, linker and debugger from within the editor such that on completion, say, of a compilation, a return is made to the editor with the source code which has been compiled loaded and with any compilation errors flagged. There will be editor commands which will enable jumps to be made to line which are flagged as containing errors.

An important utility tool is the debugger. Modern systems offer a range of debugging options. For example, the Logitech Modula system provides two debuggers: a symbolic post-mortem debugger and a symbolic run-time debugger. The run-time debugger provides a number of options: single stepping, i.e., the code is executed statement by statement with execution suspended after each statement; use of breakpoints; stop at next procedure call; and stop on return from a procedure call. Both debuggers provide five different windows onto the system (only one window can be active at any one time). These are as follows.

1. Call window   This displays the chain of procedure calls up to either the point at which a run-time error occurred or the current execution point.
2. Module window   This displays the list of modules which constitute the program being run; a module can be selected in order to display its source text or data.
3. Data window   This displays the data of the procedure or module that has been selected in the Call or Module window.
4. Text window   This displays the source text of the procedure or module selected from the Call or Module window.
5. Raw window   This displays the memory image recorded when the program stopped. Information can be selected for particular addresses; the information is displayed in hexadecimal form.

Symbolic debuggers can greatly speed up development and the benefits of an integrated system in which a return is made directly from the debugger to the editor can be seen during the final stages of development, for it is during this period that the cycle of edit, compile, load, run and debug is repeated very frequently.

For the construction of large system utilities such as version control, cross-reference generators and data-base managers become important. For example, in the development of a large system a number of different versions may exist – a current test version, previous version, new modules for the next version –

and hence there will be a range of different versions of the same module. It is vital in constructing the system or part of the system for test that there is some means of controlling which versions are incorporated into the program. It is not always the most recent version which is required.

Having said all the above, the start of a major project should be considered as an opportunity to review the choice of programming language. Table 9.1 shows a comparison of several languages in terms of the characteristics described in Chapter 9. Barney has summarized the performance as follows:

CORAL 66	27.5
FORTRAN	27.5
BASIC	28.5
Pascal	32
RTL/2	33
Ada	42
Modula-2	44

The maximum score is 60 and the criteria used in making the assessment are subjective, hence others may arrive at a different score. The assessment does, however, indicate the improvement in real-time languages brought about by the development of Ada and Modula-2.

The emphasis is increasingly turning towards computer control systems involving distributed computers. Languages which support such applications include CUTLASS with its ability to organize communications between schemes running on a number of loosely coupled computer systems. The language Conic, developed by Kramer, et al. [Kramer 1983, Duce 1984] is also finding applications for systems of loosely coupled systems. A survey of languages for use in distributed systems can be found in Duce [1984]. An interesting experiment on the use of closely coupled computer systems for process control applications has been reported by Kirrmann [1984]. In this work Modula-2 was used to write the kernel for the support of the multi-processor system. It should be noted that the definition of Modula-2 makes no assumption that the resulting code will be implemented on a single processor.

REFERENCES AND BIBLIOGRAPHY

BARNES, J.G.P. (1976), *RTL/2 Design and Philosophy*, Heyden
BARNES, J.G.P. (1982), *Programming in Ada*, Addison Wesley
BARNEY, G. (1986), *Intelligent Instrumentation*, Prentice Hall
BOOCH, G. (1983), *Software Engineering with Ada*, Benjamin Cummings Pub., Menlo Park, CA
BUHR, R. (1984), *System Design with Ada*, Prentice Hall
BRADLAW, H.S. (1982), 'An undergraduate course in real-time computer systems', *Int. J. Elect. Eng. Educ.* **19(4)**, pp. 367–77
BRANSBY, M.L. (1984), 'DDC in CEGB power stations', in Bennett, S., Linkens, D.A., (eds.) *Real-time Computer Control*, Peter Peregrinus

Table 9.1 Language comparisons: language requirements

User requirements	RTL/2	CORAL 66	Pascal	MODULA-2	ADA
Data typing	Weak	Weak	Strong not well implemented	Strong but many omissions	Very strong
Structure	Weak	V. weak	Medium	Strong	Very strong
Multiprogramming	Not included: uses semaphores and signals	Obtained via operating system	Not included	Included but all processes are anonymous	Powerful and elegant
High level programming of non-standard I/O	Not supported requires access via operating system	Some manipulation of device registers	Not included	Provides a set of constructs	Provided: better than most, but restricted
Error handling	Powerful via GOTO, insecure	Weaker use of GOTO, more secure	Weaker use of GOTO, more secure	None	Very powerful
Programming tools:					
Separate compilation	Well defined provision	Available	Not available	Very good provision	Essential
Initialization of variables	Yes	?	No	No	Yes
High level programming of standard I/O	Not supported	None	Provided by built in features, not expressible in language	None: but can write some	Standard packages (modules) defined
Generic program units	None	None	None	None	Provided
Based on:	Algol 68	Algol 60	Algol 60	Pascal	Pascal
Type:	Engineer	Government (UK)	Computer science	Computer science	Government (US)
Comments:	Relatively secure now British Standard established in industry	Imposed rather than naturally selected by users.	Small, popular especially on micros	Used for small embedded systems Standard emerging	Large, complex Futuristic

User requirements	RTL/2		CORAL 66		Pascal		MODULA		ADA		FORTRAN		BASIC	
Security	Medium	5	V low	1	High	7	High	8	V high	10	Medium	4–5	Medium	4–5
Readability	Medium	5	V poor	1–4	V high	8	V high	8	V high(a)	8	Medium	4	Poor	3
Flexibility	V good(b)	7	V good(c)	7	Fair	4	High	8	V high(d)	9	Good	5	Poor	3
Simplicity	Good	5	Good	5	Good	5	Good	5	V low	1	Fair	4	High	9
Portability	Severe(e)	2	Fair(f)	3	V poor(g)	1	V good	7	V good(h)	8	Poor(i)	2	Poor	2
Efficiency	V high	9	V good	9	V good	7	High	8	Good	6	High	8	V good(j)	7

NOTES:
(a) Should be 10, but problems
(b) For small/large systems with MTS
(c) With MASCOT
(d) Not for small systems
(e) British Standard
(f) British Standard
(g) ISO Standard
(h) ANSI Standard
(i) ISO Standard
(j) Compiled program

BULL, G. and LEWIS, A. (1983), 'Real-time BASIC', *Software: Practice and Experience*, **13(11)**, p. 1075

DOWNES, V.A. and GOLDSACK, S.J. (1982), *Programming Embedded Systems with Ada*, Prentice Hall

DUCE, D.A. (1984), *Distributed Computing Systems Programme*, Peter Peregrinus

FEUER, A. and GEHANI, N.H. (1984), *Comparing and Assessing Programming Languages, Ada, C and Pascal*, Prentice Hall

HENRY, R. et al. (1985) *The ModOS User's Manual*, Human-Computer Interaction Group, Psychology Department, Nottingham University

KIRRMANN, H.D. and KAUFMANN, F. (1984), 'PoolPo – a pool of processors for process control applications', *IEEE Trans. on Computers*, **33**., pp. 869–78

KNEPLEY, E.D. and PLATT, R. (1985), *Modula-2 Programming*, Prentice Hall

KRAMER, J., MAGEE, J. and SLOMAN, M. (1983), 'Conic: an integrated approach to distributed computer control', *IEE Proc. E.*, **130**, pp. 1–10

LEE, R. and NICHOLS, J. (1982), 'Process control BASIC simplifies programming', *Instrum. and Control Systems*, **55(1)**, pp. 51–4

MELLICHAMP, D.A. (ed.) (1983), *Real-time Computing with Applications to Data Acquisition and Control*, Van Nostrand Reinhold

MESSER, P.A. and MARSHALL, I. (1986), *Modula-2; Constructive Program Development*, Blackwell

MORALLEE, D. (1984), 'Programming languages: Where next?' *Electronics and Power*, **30(5)**, pp. 400–5

POMBERGER, G. (1986), *Software Engineering and Modula-2*, Prentice Hall

PYLE, I.C. (1981), *The Ada Programming Language*, Prentice Hall

SANDMAYR, H. (1981), 'A comparison of languages: CORAL, Pascal, PEARL, Ada and ESL', *Computers in Industry*, **2(2)**, pp. 123–32

SMEDMA, C.H., MEDEMA, P. and BOASSON, M. (1983), *The Programming Languages Pascal, Modula, CHILL and Ada*, Prentice Hall

TENDULKAR, G.A. (1983), 'Microprocessor-based industrial controllers', in Tzafestas (1983)

TZAFESTAS, S.G. (ed.) (1983), *Microprocessors in Signal Processing, Measurement and Control*, Reidel

WIENER, R. and SINOVEC, R. (1984), *Software Engineering with Modula-2 and Ada*, John Wiley

WIRTH, N. (1986), *Programming in Modula-2*, Springer Verlag (3rd edition)

WOODWARD, P.M., WETHERALL, P.R. and GORMAN, B. (1970), *The Official Definition of CORAL 66*, HMSO

YOUNG, S.J. (1982), *Real-time Languages: Design and Development*, Ellis Horwood

YOUNG, S.J. (1983), *An Introduction to Ada*, Ellis Horwood

Index

abstract model, 145–6, 168
ACCEPT, 262–7
activity, 150–1
actuation, 6
Ada, 262–8, 283, 286, 289, 297, 300–2
addressing techniques, 51
alarms, 36, 67, 249
aliasing, 116
arbitration, 236
array bounds, 271–2
assembler, 176, 188–9, 270, 305–7
asynchronous transmission, 88–9, 94

bad data, 347–8
ballast coding, 103-6
BASIC, 272–4, 281–4, 298, 301, 308–15
    real-time, 312
batch processes, 19, 40–3
BDOS, 179, 183–7, 188, 194
binary semaphore, 242
BIOS, 179, 188, 194
blocked devices, 184–5
blocked structured software, 341–2
bootstrap loader, 179
buffer, 249–53
bumpless transfer, 106–9
bus structure, 53–4, 92
busy wait, 240, 251

channel, 151, 154, 253–4, 257
characteristic error, 283
clock interrupt, 247
clock-level scheduling, 262
code sharing, 219–22
command processing, 179–82
common memory, 238, 249
communication, 8, 43–4, 87–94, 150–1, 161, 230
computation time, 138
computer control
    centralized, 38–9
    distributed, 43–5

economics, 45–6
hierarchical, 39–43
concurrency 8, 235–7
concurrent programming, 230, 319, 324–8
condition flag, 240–3, 248, 258
constants, 288
context schema, 168–9
continuous processes, 19
control flows, 147
control queue, 253–5
control structures, 289–91
control transformation, 147–9, 165
controller realization, 120-2
CORAL, 315–6
co-routine, 303, 324–7, 335
coupling and cohesion, 133
CP/M, *see* operating systems
CPU, 48–52
critical section, 144, 239, 246, 252
CUTLASS, 308, 324–50
cyclic tasks, 204–7, 214

DARTS, 149
data dictionary, 154, 156, 168
data, flow design, 145–9, 162–3, 167
data sharing, *see* shared data
data transfer
    buffered, 225–30
    non-buffered, 225–8, 231
    techniques, 85–7
    timing, 65–6
data transformation, 145–50, 163
deadlock, 214, 252, 266
debuggers, 176
debugging, 191, 305, 354
declaration, 281–4
definition module, 296–7, 320, 325
delay, 266–7
derived types, 286
development system, 305–6
device control block, 222–4
device driver, 222–4

device interrupt, 222
device queues, 230
direct digital control (DDC), 2, 24, 26–32
direct memory access (DMA), 64, 87, 236
discretization, 100, 124

error detection, 271–2, 298, 300
error trapping, 298–9
event handling, 311
exception handling, 298–301
executive program, 194
extended memory, 218

filtering, 116–7
foreground-background system, 138, 140–4, 189–92, 329
FORTRAN, 283–5, 291–4, 296–8, 301, 312–5, 320
  real-time, 314–5
functional decomposition, 132

global variable, 248, 293, 349–50
GOTO, 289–91, 293

HDLC, 90, 93
hierarchical control, 2
high level language, 307–10
human interfacing, 36

implementation model, 145–6
implementation module, 296, 320–3
independent compilation, 296–8
information hiding, 132–3, 295
initialization, 288
input image, 7–8, 158–9
input–output devices, 183–5
input–output systems, 53, 158–9, 177, 222–30
instruction set, 51
integral saturation, 103, 109–15
interface minimization, 133
interfacing, 50
  analog, 54, 61–3
  digital, 54, 58–9
  pulse, 54, 59–61, 63
  standard, 94–5
  telemetry, 54
  timing, 55–6, 60–1
interrupt, 10, 63–4, 67–87, 143–4, 328–30
  daisy chain, 72–5
  enable/disable, 241
  external interrupt, 103
  hardware-biased, 71, 72–5
  input mechanisms, 69-71

masking, 81–5
multi-level, 80–5
priority, 77–8
response vector, 71–3, 80–2, 190
restoring registers, 68–9
saving registers, 68–9
service routine, 67, 69, 140, 191, 223–4, 302–2, 329
software-biased, 71, 75
structure, 51
Z80, 69, 72, 74, 76–8, 80–4
IOTRANSFER, 328–30

kernel, 176–7, 235, 323, 331–3
keyboard input, 228–30

language requirements, 270–1
libraries, 297
linkers, 176
loaders, 176
local area network, 92–4
local variable, 248, 293
lock, 220
LOOP ... END, 291–2, 319
low level facilities, 301–3, 305, 323–6

MASCOT, 149–62, 172–3, 176, 232, 253–7
memory allocation, 181, 215–7
memory management, 215–8
memory map, 216
memory protection, 191
message passing, 225–6, 231, 248–9
Modula-2, 150, 284–6, 288, 291–7, 302–3, 318–36
modular structure, 138–44, 234, 274, 294–6, 319–36
monitor, 176, 258–62
monitoring, 5
monitor types, 178
multi-tasking, 138, 144–5, 225, 330
mutual exclusion, 221, 230–1, 237–49, 251, 259, 330

noise, 115–8

operating systems, 15, 234, 261
  CP/M, 140, 179–88, 189, 191, 194
  multi-tasking, 191, 192–4
  multi-user, 191–2
  RT/11, 140
  single/job, 178–88
operator displays, 36
output image, 7–8, 158–9

package, 274, 337
PAISLey, 149
partitioned memory, 215–6
Pascal, 274–5, 283–8, 294, 298–300
PEEK, 301, 311
PID control algorithm, 221–2, 346
  alternative forms, 106–18
  basic form, 27, 99–101
  FORTRAN program, 102
  incremental form, 106–8
  Pascal program, 101–2
  position algorithm, 108
  simple implementation, 101–3
  tuning, 118–9
  velocity algorithm, 108–9, 112, 117–8
  z-transform form, 123–4
pointers, 287–8
POKE, 301–311
polling, 10, 64–7, 103–4, 225
pool, 151–2, 154, 161
portability, 280
pre-emption, 242
primitive, 236, 239–40, 266, 318
PROCESS, 326–8
producer-consumer, 249–53
program layout, 272
programmable controllers, 24
programming
  multi-tasking, 14
  real-time, 14, 15
  sequential, 14
program timing, 104–6
prompts, 165–7

quantization, 119
queues, 242–4, 251, 253, 261

readability, 272–5, 305, 337
real-time clock, 63–4, 67, 103–4, 106, 158, 201, 205, 210
real-time systems
  classification, 8
  clock-based, 8, 10
  definition, 11
  event-based, 8
  general model, 149
  history, 1
  interactive, 8, 11
  sensor-based, 10
  software, 270
  structured development, 149, 162–72
  type 1, 11, 137
  type 2, 11, 12, 137

re-entrant code, 219–21
rendezvous, 144, 262–7
resource allocation, 194, 237
response time, 270
RTL/2, 316–8
run-time error, 285
run-time support, 176, 298
run-time testing, 271

sampling interval, 12, 101, 103, 116, 119–20, 158
saturation, 27
scope, 291–4, 297, 348–9
SELECT, 265–7
semaphores, 231, 242–9, 251–2, 258, 331–2
separate compilation, 296–8
sequence control, 21–6
shared data, 137, 142–3, 193, 237–8, 330
signals, 153–4, 236, 251–2, 328, 333
software bus, 234–5, 323
software design
  general, 137–9
  real-time, 145–73
software modules, 131–3
software specification, 129, 133–5
spooling, 225
storage, 52–3
structured programming, 290
structured types, 287
supervisory control, 20, 24, 33–6
synchronization, 7, 102–3, 144, 152–3, 236, 251–2, 264
synchronous transmission, 90–1, 94
system calls, 185–7, 213

task
  base level, 202, 208–9
  chaining, 217–8
  clock level, 202, 204
  communication, 256
  descriptor, 197–201, 216, 221, 245–6, 324
  dispatch, 207, 210–1, 212, 241
  interrupt level, 201, 203
  management, 195–214
  overlaying, 218
  priority, 197–8, 200–2, 208, 210, 247–8, 261
  scheduling, 144, 201–9, 247, 262
  states, 195–7, 200, 209, 227, 312, 330–1
  swapping, 215, 217–8, 220
  synchronization, 213, 259, 261, 263
time out, 266
time slicing, 208

timing, 157–8, 332
token passing, 94
transaction processing, 145
transformation schema, 169–70, 173
transformation specification, 170–1
types, 281, 284–7

virtual machine, 176, 188, 280
visibility, 291–5, 348–9

volatile environment, 220

watchdog timer, 61
wind-up, *see* integral saturation
wordlength, 51, 119, 207

Ziegler and Nichols, 118
Zilog PIO, 58–9